SENSORS AND CONTROLS IN THE ANALYSIS OF DISTRIBUTED SYSTEMS

MATHEMATICS AND ITS APPLICATIONS

Series Editor: G. M. BELL, Professor of Mathematics, King's College London (KQC), University of London

STATISTICS AND OPERATIONAL RESEARCH

Editor: B. W. CONOLLY, Professor of Operational Research, Queen Mary College, University of London

Mathematics and its applications are now awe-inspiring in their scope, variety and depth. Not only is there rapid growth in pure mathematics and its applications to the traditional fields of the physical sciences, engineering and statistics, but new fields of application are emerging in biology, ecology and social organization. The user of mathematics must assimilate subtle new techniques and also learn to handle the great power of the computer efficiently and economically.

The need for clear, concise and authoritative texts is thus greater than ever and our series will endeavour to supply this need. It aims to be comprehensive and yet flexible. Works surveying recent research will introduce new areas and up-to-date mathematical methods. Undergraduate texts on established topics will stimulate student interest by including applications relevant at the present day. The series will also include selected volumes of lecture notes which will enable certain important topics to be presented earlier than would otherwise be possible.

In all these ways it is hoped to render a valuable service to those who learn, teach, develop and use mathematics.

Mathematics and its Applications

Series Editor: G. M. BELL, Professor of Mathematics, King's College London (KQC), University of London

Artmann, B.	**The Concept of Number**
Arczewski, K. & Pietrucha, J.	**Mathematical Modelling in Discrete Mechanical Systems**
Arczewski, K. and Pietrucha, J.	**Mathematical Modelling in Continuous Mechanical Systems**
Bainov, D.D. & Konstantinov, M.	**The Averaging Method and its Applications**
Balcerzyk, S. & Joszefiak, T.	**Commutative Rings**
Balcerzyk, S. & Joszefiak, T.	**Noetherian and Krull Rings**
Baldock, G.R. & Bridgeman, T.	**Mathematical Theory of Wave Motion**
Ball, M.A.	**Mathematics in the Social and Life Sciences: Theories, Models and Methods**
de Barra, G.	**Measure Theory and Integration**
Bell, G.M. and Lavis, D.A.	**Co-operative Phenomena in Lattice Models, Vols. I & II**
Berkshire, F.H.	**Mountain and Lee Waves**
Berry, J.S., Burghes, D.N., Huntley, I.D., James, D.J.G. & Moscardini, A.O.	**Mathematical Modelling Courses**
Berry, J.S., Burghes, D.N., Huntley, I.D., James, D.J.G. & Moscardini, A.O.	**Mathematical Methodology, Models and Micros**
Berry, J.S., Burghes, D.N., Huntley, I.D., James, D.J.G. & Moscardini, A.O.	**Teaching and Applying Mathematical Modelling**
Brown, R.	**Topology**
Burghes, D.N. & Borrie, M.	**Modelling with Differential Equations**
Burghes, D.N. & Downs, A.M.	**Modern Introduction to Classical Mechanics and Control**
Burghes, D.N. & Graham, A.	**Introduction to Control Theory, including Optimal Control**
Burghes, D.N., Huntley, I. & McDonald, J.	**Applying Mathematics**
Burghes, D.N. & Wood, A.D.	**Mathematical Models in the Social, Management and Life Sciences**
Butkovskiy, A.G.	**Green's Functions and Transfer Functions Handbook**
Butkovskiy, A.G.	**Structural Theory of Distributed Systems**
Cao, Z.-Q., Kim, K.H. & Roush, F.W.	**Incline Algebra and Applications**
Cerny, I.	**Complex Domain Analysis**
Chorlton, F.	**Textbook of Dynamics, 2nd Edition**
Chorlton, F.	**Vector and Tensor Methods**
Cohen, D.E.	**Computability and Logic**
Crapper, G.D.	**Introduction to Water Waves**
Cross, M. & Moscardini, A.O.	**Learning the Art of Mathematical Modelling**
Cullen, M.R.	**Linear Models in Biology**
Dunning-Davies, J.	**Mathematical Methods for Mathematicians, Physical Scientists and Engineers**
Eason, G., Coles, C.W. & Gettinby, G.	**Mathematics and Statistics for the Bio-sciences**
El Jai, A. & Pritchard, A.J.	**Sensors and Controls in the Analysis of Distributed Systems**
Exton, H.	**Handbook of Hypergeometric Integrals**
Exton, H.	**Multiple Hypergeometric Functions and Applications**
Exton, H.	***q*-Hypergeometric Functions and Applications**

Series continued at back of book

SENSORS AND CONTROLS IN THE ANALYSIS OF DISTRIBUTED SYSTEMS

A. EL JAI
Professor, Faculté des Sciences, University of Rabat, Morocco
and Laboratoire d'Automatique, University of Perpignan, France
and
A. J. PRITCHARD
Professor, Department of Mathematics, and Director of Control Theory Centre
University of Warwick

ELLIS HORWOOD LIMITED
Publishers · Chichester

Halsted Press: a division of
JOHN WILEY & SONS
New York · Chichester · Brisbane · Toronto

This English Edition first published in 1988 by
ELLIS HORWOOD LIMITED
Market Cross House, Cooper Street,
Chichester, West Sussex, PO19 1EB, England
The publisher's colophon is reproduced from James Gillison's drawing of the ancient Market Cross, Chichester.

Distributors:
Australia and New Zealand:
JACARANDA WILEY LIMITED
GPO Box 859, Brisbane, Queensland 4001, Australia
Canada:
JOHN WILEY & SONS CANADA LIMITED
22 Worcester Road, Rexdale, Ontario, Canada
Europe and Africa:
JOHN WILEY & SONS LIMITED
Baffins Lane, Chichester, West Sussex, England
North and South America and the rest of the world:
Halsted Press: a division of
JOHN WILEY & SONS
605 Third Avenue, New York, NY 10158, USA
South-East Asia
JOHN WILEY & SONS (SEA) PTE LIMITED
37 Jalan Pemimpin # 05–04
Block B, Union Industrial Building, Singapore 2057
Indian Subcontinent
WILEY EASTERN LIMITED
4835/24 Ansari Road
Daryaganj
New Delhi 110002, India

Translated by: Catrin Pritchard, BSc (Oxon), PhD, Imperial Cancer Research Fund, London and Rhian Pritchard, BSc (Oxon), Department of Physics, Imperial College of Science and Technology, London.
First published in French by Masson Editeur, Paris, under the title *Capteurs et actionneurs dans l'analyse des systémes distribués*

British Library Cataloguing in Publication Data
El Jai, A.
Sensors and controls in the analysis of distributed systems. —
(Mathematics and its applications/Ellis Horwood).
1. Control theory 2. Distributed parameter systems
I. Title II. Pritchard, A. J.
III. Capteurs et actionneurs dans l'analyse des systems distribues. *English* IV. Series
003 QA402.3

Library of Congress CIP Data available

ISBN 0–7458–0336–9 (Ellis Horwood Limited)
ISBN 0–470–21023–0 (Halsted Press)

Printed in Great Britain by Unwin Bros., Woking

Table of contents

Preface

The object of this book is to present the principal results in the analysis of distributed parameter systems which are concerned with the structure of the input and output operators (i.e. controls and sensors).

We have adopted an approach via semigroups since this is natural for infinite dimensional systems. Various ideas and results have been extended from finite dimensional theory and, in so doing, we have isolated many notions which require rather delicate manipulations; we have chosen a mathematical style which allows us to consider problems in a rigorous fashion and have obtained results for cases of practical interest, but we do not consider the most general situations. Where the proofs of the results are constructive or original, we have presented them. The proofs which require some rather sophisticated functional analysis are not given.

We hope that the book will be of interest to the automatician, applied mathematician and engineer.

We would like to thank Professor J. L. Lions for a number of helpful comments which have introduced new ideas and a clarification and enhancement of some of the results, the members of the 'optimal control' team of the Department of Mathematics and Computer Science, Faculty of Science, Rabat, for the development of both numerical simulations and the general theory of this book, and Catrin and Rhian Pritchard for the translation from French to English.

Notation

GENERAL NOTATION

Linear operators

$D(H)$	domain of H
$\mathrm{Im}(H)$	image of H
$\mathrm{Ker}(H)$	kernel of H
H^*	adjoint of H
H^{T}	transpose of H
$H^{\dagger}$	pseudo-inverse of H
$\sigma(H)$	spectrum of H
I	identity operator
$\mathrm{P}_A x$	projection of x on A
$\mathrm{rank}(H)$	rank of the matrix H
$\langle\ ,\ \rangle$	scalar product in duality

Set notation

$\bar{A}$	closure of A
$\mathring{A}$	interior of A
$A^{\perp}$	orthogonal of A
$\varnothing$	empty set
$\mathrm{Re}(z)$	real part of z
Ω	open bounded set in $\mathbb{R}^n$ with smooth boundary
$\partial\Omega$	boundary of Ω
$\mathbb{Q}$	rational numbers
$\mathrm{supp}(g)$	support of the function g

Functions

$f(.)$	function $x \to f(x)$
$\mathscr{F} f$	Fourier transform of f
Δf	Laplacian of $f = \sum_{i=1}^{n} (\partial^2 f / \partial x_i^2)$
$\partial f / \partial v$	exterior normal derivative of f

Function spaces

$\mathscr{L}(X, Y)$	the space of bounded linear maps from X to Y
$\mathscr{L}(X)$	$\mathscr{L}(X, X)$
X'	dual space of $X = \mathscr{L}(X, \mathbb{R} \text{ or } \mathbb{C})$
$L^r(0, T; E)$	integrable functions $f:]0, T[\to E$ such that $t \to \| f(t) \|^r$ is integrable on $]0, T[$
$L^2(\Omega)$	space of functions square integrable on Ω
$H^m(\Omega)$	Sobolev space of order $m = \{ f \in L^2(\Omega): D^\alpha f \in L^2(\Omega) \}$ for all $\alpha = (\alpha_1, \ldots, \alpha_n) \in \mathbb{N}^n$ such that $\lvert\alpha\rvert = \alpha_1 + \cdots + \alpha_n \leqslant m$ and $D^\alpha = D_1^{\alpha_1} \cdots D_n^{\alpha_n}$, $D_j = \partial / \partial r_j$
$H_0^m(\Omega)$	$\{ f \in H^m(\Omega): \partial^k f / \partial v^k \rvert_{\partial\Omega} = 0;\ 0 \leqslant k \leqslant m-1 \}$
$H^{-m}(\Omega)$	dual of $H_0^m(\Omega)$
$\mathscr{D}(0, T)$	space of infinitely differentiable functions with compact support in Ω

PARTICULAR NOTATION

(S)	system represented by $\dot{z} = Az + Bu$, $z(0) = z_0$
(E)	output equation $y = Cz$
(S_z)	diffusion system excited by zone controls
(S_p)	diffusion system excited by point controls
(S_f)	diffusion system excited by zone controls on the boundary
(S_f')	diffusion system excited by point controls on the boundary
$\mathscr{E}, \mathscr{U}, \mathscr{O}$	state, control and output spaces
$A, B, \ldots$	operators
■	end of a proof

Some equations are not presented in a homogeneous manner since we wish to emphasize the precise dependence of the functions on space and time variables.

The systems considered are governed by partial differential equations which we sometimes call distributed, or distributed parameter systems.

Introduction

Many real systems are intrinsically of a distributed nature and many systems constructed by man are very complex. As a result, the systems are difficult to model and the equations describing the systems are difficult to solve. Although the mathematicians have developed sophisticated methods to do this, the automatician has a tendency to use only empirical methods which nevertheless are sometimes very effective.

If, today, few distributed parameter system models are used by the automatician, it is essentially because of the mathematical difficulties confronted in moving from finite to infinite dimensions. However, how can one ignore the distributed nature of heat flow in a furnace? How can one ignore the flexible nature of large space structures in stabilization problems? Coarse approximation schemes may be alright for some applications but, in many cases, one should not neglect the distributed nature and more sophisticated tools must be used. In any case, in recent works on control, one finds an introduction to distributed systems and it is sure that this tendency will continue and evolve.

The problems of the control of distributed systems have been the subject of a good deal of research. In the deterministic case, J. L. Lions has played a leading role and developed much of the theory. The literature is vast, with contributions from, amongst others, Balakrishnan, Butkovskii, Curtain, Delfour, Fattorini, Dolecki, Lasiecka, Mitter, Pandolfi, Pritchard, Russell, Sakawa, Slemrod, Triggiani, Wang and Zabczyk.

In this book, we concentrate on the effect that the structure of controls and sensors has on the analysis of certain classes of distributed systems. Spatial variations play an important part. For example if one studies the controllability of the system

$$(S) \qquad \begin{cases} \dot{z} = Az + Bu \\ z(0) = z_0 \end{cases}$$

one needs to consider the pair (A, B). Both of these operators depend on spatial distributions, the first on the region over which the system is modelled and the second on the distribution of the control.

It is possible to study the effect that the structure of controls and sensors have on many systems, continuous or discrete, deterministic or stochastic. For example in a diffusion furnace the sensors are thermocouples, and the controls are the burners. In traffic flows the controls may be the traffic lights and the sensors may furnish information about the density and speed of the traffic. Our study is essentially limited to systems described by parabolic partial differential equations although we give some consideration to hyperbolic and non-linear systems.

It may seen rather premature to analyse the structure of controls and sensors since in general these are given quantities. However, there is a tendency to make systems as perfect as possible and this includes optimizing the parameters of the structure of the sensors and controls. For a flexible work station or a configuration of ovens and burners, optimum design is not usually attempted. However, we shall see that choice of sensors and controls is very important and the parameters which define them can be made the subject of optimization studies.

In this work the models that we choose are more or less academic and the results that we obtain are specific to them. Real problems are usually much more complex, non-linear, etc., and would require a much more concentrated individual treatment. Nevertheless, we hope to illustrate the types of problem that can be posed.

Chapter 1 is concerned with general principles for the analysis of distributed parameter systems. More precisely, we introduce the notions of exact and weak controllability and then so-called strategic controls. By duality, we make a similar study of observability and strategic sensors. We also introduce the notion of stabilizability and the dual notion of detectability. Then we show how it is possible to design compensators.

Most of the work in Chapter 2 is concerned with parabolic systems with some standard geometric configurations. We then analyse the effect of the structure of various types of control and sensor on the problems of controllability and observability.

Chapter 3 is concerned with optimizing the structure of the controls. We again only consider this for parabolic systems.

Finally, in Chapter 4, we present some results for hyperbolic and non-linear systems. We show that, with some extra analysis, it is possible to develop theories of controllability and observability similar to those for linear parabolic systems.

Chapter 1
General principles

Let us consider the system described by the abstract differential equation

$$(S) \qquad \begin{cases} \dot{z} = Az + Bu & 0 < t < T \\ z(0) = z_0 \end{cases}$$

This formulation is chosen since it generalizes the classical finite dimensional state space representation. Mathematically, it is very general and allows us to describe many of the ideas in the analysis.

We suppose that the system (*S*) is augmented by an output equation

$$(E) \qquad y = Cz$$

We shall study various ideas in the analysis of (*S*) and (*E*) associated with the choice of the operators *B* and *C*, that is to say, the different ways that the input can excite the system (*S*) and the different ways that information can be retrieved about the system through (*E*). For example, in the case of *B*, if Ω is the geometric domain of a physical system modelled by (*S*), we shall examine in this book the following types of input action:

(1) zones, over a part of Ω or its boundary;
(2) points, in Ω or its boundary;
(3) filaments, in Ω or its boundary.

Mathematically, this corresponds to the case where the support of the input is of dimension

(1) equal to that of Ω or that of the boundary for zones;
(2) equal to zero for points;
(3) equal to one for filaments.

Similarly, we shall examine various types of support for the output equation (*E*).

The problems that we shall consider are those of controllability and stabilizability (and their link with the choice of B) and the dual notions of observability and detectability (and their link with the choice of C).

Remark 1.1

It is also possible to consider filaments with supports of higher dimension than one.

Hypotheses

Without loss of generality we shall take $z_0 = 0$ and suppose for the moment that we have the following hypotheses.

Hypothesis 1.1

$\mathscr{E}, \mathscr{U}$ and $\mathscr{O}$ are separable Hilbert spaces called the state, input and output spaces

Hypothesis 1.2

$$u \in L^2(0, T; \mathscr{U}), \quad B \in \mathscr{L}(\mathscr{U}, \mathscr{E}) \quad \text{and} \quad C \in \mathscr{L}(\mathscr{E}, \mathscr{O}).$$

Hypothesis 1.3

A is self-adjoint with compact resolvent and generates a strongly continuous semigroup $(S(t))_{t \geq 0}$ on $\mathscr{E}$. ■

Under these hypotheses, (S) has a unique weak solution, strongly continuous on $[0, T]$ and given by

$$z(t) = \int_0^t S(t-\tau) B u(\tau) \, \mathrm{d}\tau \tag{1.1}$$

We denote by H the bounded linear operator

$$u \in L^2(0, T; \mathscr{U}) \to Hu \in \mathscr{E}$$

defined by

$$Hu = \int_0^T S(T-\tau) B u(\tau) \, \mathrm{d}\tau \tag{1.2}$$

Remark 1.2

(1) The hypotheses on B and C are very strong. They will be modified for the cases of point and boundary control.
(2) Some of the results which we develop can be extended to the case where A depends on time.
(3) For controllability and observability problems, we shall often take $Az = \Delta z$ whereas, for stabilizability and detectability problems, $Az = \Delta z + kz$. In both cases, A is elliptic and the regularity of the solution depends on the term Bu (Dautry and Lions, 1984).

(4) Hypothesis 1.3 implies that A has a denumerable discrete spectrum

$$\sigma(A) = \{\lambda_1, \lambda_2, \ldots, \lambda_k, \ldots\}$$

with

$$\lim_{k \to +\infty} (\lambda_k) = -\infty$$

(5) In many of the situations, we shall have

$$\mathscr{E} = L^2(\Omega), \qquad \mathscr{U} = \mathbb{R}^p, \qquad \mathscr{O} = \mathbb{R}^q$$

(p and q are determined by the numbers of actuators and sensors).

Semigroups

The family $(S(t))_{t \geqslant 0}$ of operators in $\mathscr{L}(\mathscr{E})$ have the properties

$$\begin{array}{lll} (\alpha) & S(0) = I & \\ (\beta) & S(t)S(s) = S(s)S(t) = S(t+s) & \forall s, t \geqslant 0 \\ (\gamma) & \| S(t)z - z \| \to 0 \quad \text{as } t \to 0^+ & \forall z \in \mathscr{E} \end{array}$$

The operator A defined by

$$Az = \lim_{t \to 0+} \left\{ \frac{S(t)z - z}{t} \right\}$$

is the infinitesimal generator of the semigroup $(S(t))_{t \geqslant 0}$. Its domain is the set of $z \in \mathscr{E}$ such that the above limit exists.

A necessary and sufficient condition for a closed densely defined linear operator A to generate a strongly continuous semigroup is that there exists real numbers M, λ such that for all $\xi > \lambda$:

$$\| (\xi I - A)^{-k} \| \leqslant \frac{M}{(\xi - \lambda)^k} \qquad k = 1, 2, \ldots$$

If A has a complete orthonormal set of eigenfunctions (ϕ_{nj}) with associated eigenvalues (λ_n), λ_n having multiplicity r_n, then the semigroup generated by A has the form

$$S(t)z = \sum_n e^{\lambda_n t} \sum_{j=1}^{r_n} \langle z, \phi_{nj} \rangle \phi_{nj} \tag{1.3}$$

In the above definition, $\mathscr{E}$ need not be a Hilbert space. If $\mathscr{E}$ is a reflexive Banach space, then the adjoint A^* will generate a strongly continuous semigroup $(S^*(t))_{t \geqslant 0}$ on the dual space $\mathscr{E}'$. For more details, see Hille and Phillips (1957), Butzer and Berens (1967) and Yosida (1980).

1.1 CONTROLLABILITY

In the case of finite dimensional systems the concept of controllability is equivalent to reaching the whole of the state space from a given initial state. The essential difference in the case of distributed parameter systems is that, in general, not all the state space $\mathscr{E}$ can be reached. This is the case, for example, when the operator A is not bounded but the operator B is bounded. However, in these cases, it might be possible to reach points arbitrarily close to any state. So we introduce various degrees of controllability.

1.1.1 Exact and weak controllability

Definition 1.1 Exact controllability in $\mathscr{E}$

The system (S) is said to be exactly controllable on $[0, T]$ if

$$\forall z_d \in \mathscr{E} \qquad \exists u \in L^2(0, T; \mathscr{U}) \qquad \text{such that } z(T) = z_d \tag{1.4}$$

where $z(.)$ is defined by equation (1.1).

Remark 1.3

For the operator H defined in equation (1.2). Definition 1.1 is equivalent to

$$\mathrm{Im}(H) = \mathscr{E} \tag{1.5}$$

Definition 1.2 Exact controllability in $\mathscr{E}_1 \subset \mathscr{E}$

If $\mathscr{E}_1$ is a subspace of $\mathscr{E}$, the system (S) is said to be exactly controllable in $\mathscr{E}_1$ on $[0, T]$ if

$$\forall z_d \in \mathscr{E}_1 \qquad \exists u \in L^2(0, T; \mathscr{U}) \qquad \text{such that } z(T) = z_d \tag{1.6}$$

Remark 1.4

Definition 1.2 is equivalent to

$$\mathrm{Im}(H) \supset \mathscr{E}_1 \tag{1.7}$$

Definition 1.3 Weak controllability

The system (S) is said to be weakly controllable on $[0, T]$ if, for all $z_d \in \mathscr{E}$,

$$\forall \varepsilon > 0 \qquad \exists u \in L^2(0, T; \mathscr{U}) \qquad \text{such that } \| z(T) - z_d \| \leqslant \varepsilon \tag{1.8}$$

Remark 1.5

(1) It may be that the system is weakly controllable on $\mathscr{E}$ and exactly controllable to a subspace $\mathscr{E}_1$ (see Example 1.1).
(2) The above definition means that, although not all the states can be reached, they can be approximated arbitrarily closely.
(3) It may be possible to choose $u(.)$ in a more regular space if the state z_d is more regular.
(4) We shall say loosely that the system (S) is controllable or the pair (A, B) is controllable.

1.1.2 Characterizations

The following characterizations result from Definitions 1.1 and 1.3.

Property 1.1 Characterization of exact controllability

The system (S) is exactly controllable in $\mathscr{E}$ on $[0, T]$ if and only if

$$\exists \gamma > 0 \qquad \text{such that } \| z^* \|_{\mathscr{E}'} \leqslant \gamma \| B^* S^*(.) z^* \|_{L^2(0, T; \mathscr{U}')} \tag{1.9}$$

for all $z^* \in \mathscr{E}'$.

Proof This follows from a more general result (Curtain and Pritchard, 1978). If $\mathscr{E}_1$ and $\mathscr{U}_1$ are two reflexive Banach spaces, and $H_1 \in \mathscr{L}(\mathscr{U}_1, \mathscr{E}_1)$ then there is an equivalence between the following.

(1) $\mathscr{E}_1 \subset \mathrm{Im}(H_1)$.

There exists

(2) $\gamma > 0$ such that $\| z^* \|_{\mathscr{E}'_1} \leqslant \gamma \| H_1^* z^* \|_{\mathscr{U}'_1}$ for all $z^* \in \mathscr{E}'_1$.

Suppose then that

$$\mathscr{U}_1 = L^2(0, T; \mathscr{U}), \qquad \mathscr{E}_1 = \mathscr{E}, \qquad H_1 = H$$

Then we have that $\mathscr{U}'_1 = L^2(0, T; \mathscr{U}')$, that $(S^*(t))_{t \geqslant 0}$ is a strongly continuous semigroup and that

$$\langle z^*, Hu \rangle = \langle z^*, \int_0^T S(T-\tau)Bu(\tau)\,\mathrm{d}\tau \rangle$$

Hence,

$$\langle z^*, Hu \rangle = \int_0^T \langle B^* S^*(T-\tau) z^*, \quad u(\tau) \rangle \,\mathrm{d}\tau$$
$$= \langle H^* z^*, u \rangle$$

and so

$$H^* z^* = B^* S^*(.) z^*$$

which proves the result. ■

The above proposition is useful since it characterizes exact controllability in terms of an inequality which may be easier to compute for a given system.

We can extend the characterization to exact controllability in $\mathscr{E}_1 \subset \mathscr{E}$ by replacing the left-hand side in equation (1.9) by $\| z^* \|_{\mathscr{E}'_1}$.

Sometimes, it is possible to use other methods to determine whether or not a system is exactly controllable.

Proposition 1.1

Suppose that H_t is a linear operator $L^2(0, T; \mathscr{U}) \to \mathscr{E}$ defined by

$$H_t u = \int_0^t S(t-\tau)Bu(\tau)\,\mathrm{d}\tau$$

If, for all $t \geqslant 0$, H_t is compact, then the system is not exactly controllable.

Proof (S) is exactly controllable on $[0, T]$ if and only if

$$\bigcup_{n \in \mathbb{N}} H_T(B_n) = \mathscr{E}$$

where B_n is the ball centred at the origin and radius n in $\mathscr{U}$. This implies that

$$\bigcup_{n \in \mathbb{N}} \overline{H_T(B_n)} = \mathscr{E}$$

However, the fact that $\overline{H_T(B_n)}$ is compact implies that $\overset{\circ}{\overline{H_T(B_n)}} = \varnothing$ for all n in $\mathbb{N}^*$; so the result follows by an application of Baire's theorem. ■

Corollary

If $S(t)$ is compact, and H_t is bounded for all $t > 0$, then the system (S) is never exactly controllable.

Proof Suppose that $T_\varepsilon = T - \varepsilon$, it is easy to see that

$$S(\varepsilon)H_{T_\varepsilon}u + \int_{T_\varepsilon}^{T} S(T-\tau)Bu(\tau) \quad d\tau = Hu$$

where H is defined in equation (1.2). The facts that $S(t)$ is compact for all $t > 0$ and that H_{T_ε} is bounded implies that $S(\varepsilon)H_{T_\varepsilon}$ is compact. When $\varepsilon \to 0$, $S(\varepsilon)H_{T_\varepsilon}$ converges uniformly to H, which implies that H is compact. So the result follows by applying Proposition 1.1. ■

Corollary

If B is compact, then the system (S) is never exactly controllable.

Property 1.2 Characterization of weak controllability

For distributed parameter systems the concept of weak controllability is more useful. There is an equivalence between the following.

(1) (S) is weakly controllable on $[0, T]$.
(2) $\overline{\text{Im}(H)} = \mathscr{E}$.
(3) $\text{Ker}(H^*) = \text{Ker}(H^*H) = \{0\}$.
(4) $B^*S^*(s)z = 0$, for all $s \in [0, T]$ implies that $z = 0$.
(5) If the semigroup $(S(t))_{t \geq 0}$ is analytic, then

$$\overline{\bigcup_{n\in\mathbb{N}} \text{Im}\{A^n S(s)B\}} = \mathscr{E} \qquad \forall s \in]0, T]$$

Proof (2) and (3) are clearly equivalent to (1), and (4) follows since $(H^*z)(s) = B^*S^*(s)z$. To prove (4) implies (5). If this were not the case, there would exist $z \neq 0$ and $s \in]0, T[$ such that

$$\langle z, A^n S(s)Bv \rangle = 0 \qquad \forall n \in \mathbb{N} \quad \text{and} \quad \forall v \in \mathscr{U}$$

Now

$$\langle z, A^n S(s)Bv \rangle = \frac{d^n}{ds^n} \langle z, S(s)Bv \rangle \qquad \forall n \in \mathbb{N}$$

and, by analyticity, one deduces that

$$\langle z, S(s)Bv \rangle = 0 \qquad \forall v \in \mathscr{U} \quad \text{and} \quad s \in]0, T[$$

and this contradicts (4).

To show (5) implies (4). If this were not the case, there would exist z such that

$$\langle z, S(t)Bv \rangle = 0 \qquad \forall t \in [0, T] \quad \text{and} \quad \forall v \in \mathscr{U}$$

but then, for all $n \in \mathbb{N}$, for all $v \in \mathscr{U}$ and for all $t \in]0, T]$,

$$\frac{d^n}{dt^n} \langle z, S(t)Bv \rangle = \langle z, A^n S(t)Bv \rangle = 0$$

and so

$$\bigcup_n A^n S(t) B\mathscr{U}$$

cannot be dense in $\mathscr{E}$. ■

Example 1.1

Suppose that Ω is a bounded open set in $\mathbb{R}^n$ with a smooth boundary $\partial\Omega$, and consider the system

$$\begin{aligned} &\frac{\partial y}{\partial t} = \Delta y + u && \text{in } \Omega \times]0, T[\\ &y(x, 0) = 0 && \text{in } \Omega \\ &y(\xi, t) = 0 && \text{in } \partial\Omega \times]0, T[\end{aligned} \tag{1.10}$$

The above system is a special case of the system (*S*) if we take

$$\mathscr{E} = L^2(\Omega), \qquad Az = \Delta z$$

where

$$z \in D(A) = H^2(\Omega) \cap H_0^1(\Omega)$$
$$u \in L^2(0, T; L^2(\Omega))$$

In this case, $S(t)$ is compact for $t > 0$ and so, as a result of Corollary 1.1, the system cannot be exactly controllable in $L^2(\Omega)$. Actually, one may write the inequality (1.9) and see that this cannot be satisfied when one takes $\mathscr{E} = L^2(\Omega)$ (Bensoussan, 1972; Curtain and Pritchard, 1978). However, we show below that the system is exactly controllable to $\mathscr{E}_1 = H_0^1(\Omega)$. If (ϕ_{nj}) is a system of complete orthonormal eigenfunctions for Δ with Dirichlet boundary conditions, and (λ_n) is the corresponding set of eigenvalues, λ_n having multiplicity r_n, then

$$S(t)z = \sum_n e^{\lambda_n t} \sum_{j=1}^{r_n} \langle z, \phi_{nj} \rangle \phi_{nj}$$

Now A is self-adjoint and $B = B^* = I$, and so

$$\begin{aligned} \| B^*S^*(.)z \|^2_{L^2(0,T;L^2(\Omega))} &= \sum_n \frac{1 - e^{2\lambda_n T}}{2|\lambda_n|} \sum_{j=1}^{r_n} \langle z, \phi_{nj} \rangle^2 \\ &\geqslant \frac{1 - e^{2\lambda_0 T}}{2} \sum_n \sum_{j=1}^{r_n} \left\langle z, \frac{\phi_{nj}}{(-\lambda_n)^{1/2}} \right\rangle^2 \end{aligned}$$

but, since $(\phi_{nj}/(-\lambda_n)^{1/2})$ is a complete orthonormal set in $H^{-1}(\Omega)$, then

$$\| B^*S^*(.)z \|^2_{L^2(0,T;L^2(\Omega))} \geqslant \gamma \| z \|^2 H^{-1}(\Omega)$$

where

$$\gamma = \left(\frac{1 - e^{2\lambda_0 T}}{2} \right)^{-1/2}$$

So the result follows since $(H_0^1(\Omega))' = H^{-1}(\Omega)$. ■

The example shows that, for a diffusion system, a control distributed over all the set Ω is not sufficient to obtain exact controllability in the space $L^2(\Omega)$, but there is exact controllability to the subspace $H_0^1(\Omega)$. This explains, to some extent, why we choose to develop the concept of weak controllability in this book.

1.2 ACTUATORS

The sensors and actuators are the intermediaries between a real system and its environment. The sensors play a passive role and enable one to gather information on the system and its evolution; these will be studied in a later chapter. The actuators permit the excitation of the system, and can be of many different forms. Those most frequently met in practice are as follows.

(1) *Fixed point.* An example is a burner in a diffusion system (Fig. 1.1).
(2) *Mobile point.* This is a point actuator where the position varies in time; an example is a system excited by a moving laser.
(3) *Zone.* An example is a diffusion process with heating zones (Fig. 1.2).
(4) *Filament.* An example is an oven heated by an electric resistance (Fig. 1.3).

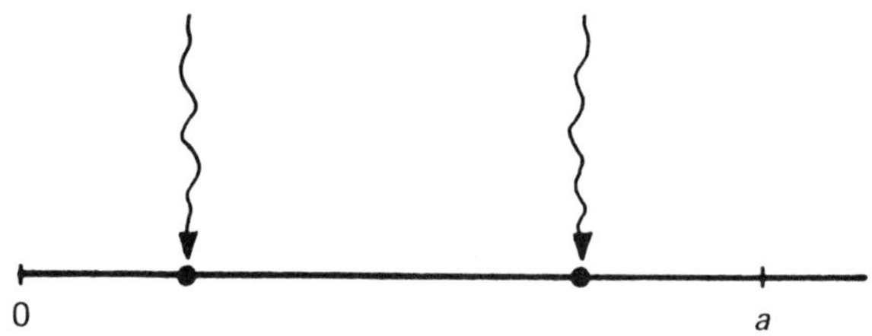

Fig. 1.1 One-dimensional system with two point actuators

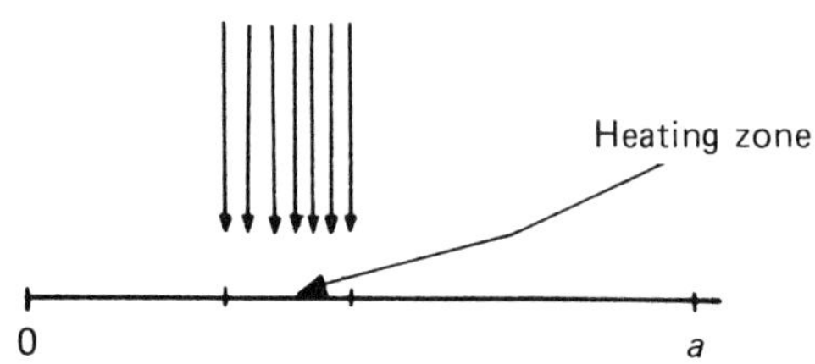

Fig. 1.2 One-dimensional system with a zone actuator

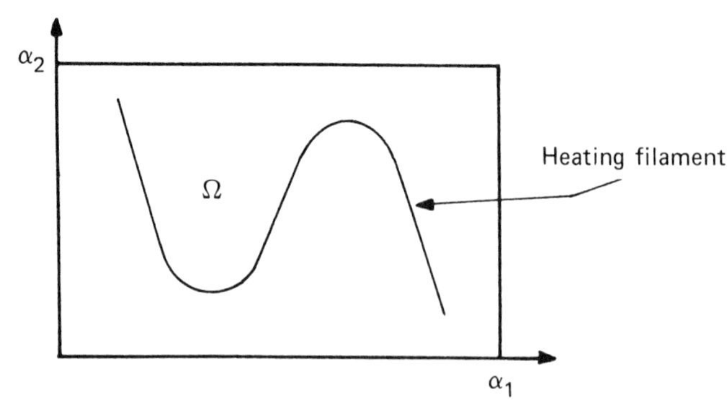

Fig. 1.3 Two-dimensional system with a filament actuator

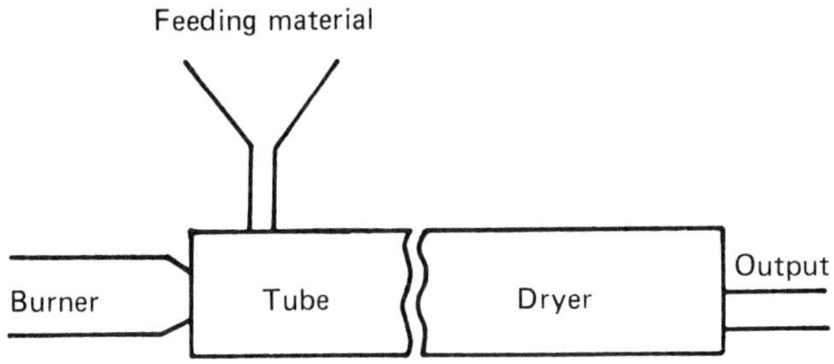

Fig. 1.4 Diagram of an oven for drying phosphates

The above types of actuator can be located in the interior of the domain Ω or on its boundary. Fig. 1.4 represents a rotating oven which is used for the drying of phosphates and which can only be excited at the extremities (Najim *et al.* 1976; Redouane, 1980).

In this book, we shall examine the case of systems excited by a finite number of zones $\Omega_i(\Omega_i \subset \Omega)$ or $\Gamma_i(\Gamma_i \subset \partial\Omega)$ or points $b_i(b_i \in \Omega$ or $b_i \in \partial\Omega)$. The case of filament actuators may be included in point actuators. For zone actuators, we shall always assume that the support Ω_i (or Γ_i) has a non-empty interior.

1.2.1 Definitions

Definition 1.4

Suppose that Ω_i is closed in Ω and $g_i \in L^2(\Omega_i)$. An actuator zone is the couple (Ω_i, g_i) where Ω_i is the support of the actuator and g_i defines the spatial distribution of the actuator.

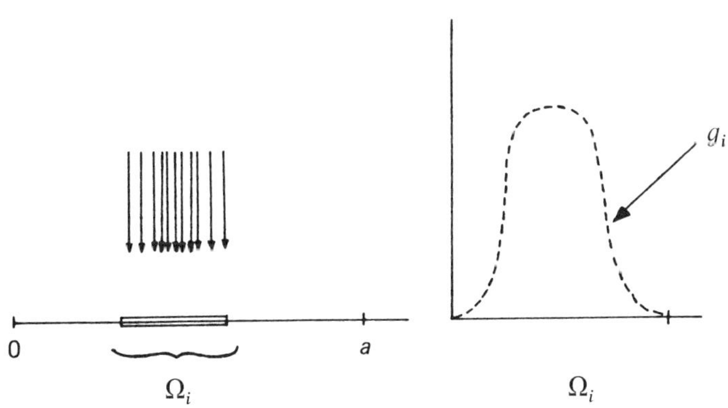

Fig. 1.5 Support and spatial distribution of (Ω_i, g_i)

Definition 1.5

Suppose that $b_i \in \Omega$; a point actuator is the couple (b_i, δ_{b_i}), where δ_{b_i} is the Dirac mass at the point b_i.

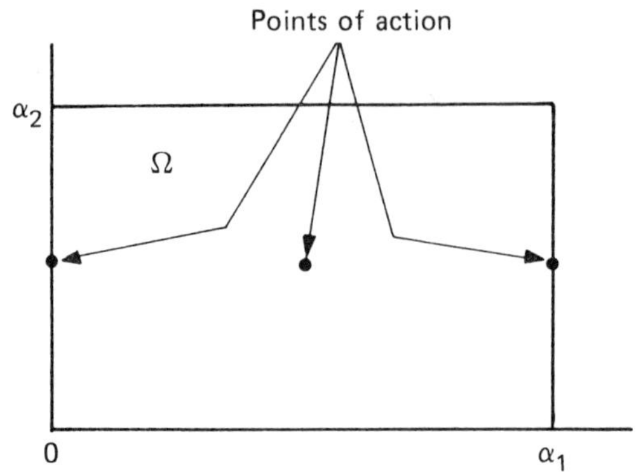

Fig. 1.6 Point controls

Remark 1.6

In the case of boundary control the definitions are the same. We say that boundary zone control is

$$(\Gamma_i, g_i), \qquad \Gamma_i \subset \partial\Omega, \qquad g_i \in L^2(\Gamma_i)$$

and that boundary point control is

$$(b_i, \delta_{b_i}), \qquad b_i \in \partial\Omega$$

Our main interest is the study of the relationship between the controls and the notion of controllability, but first we introduce the various systems which will be considered.

In the following, Ω is open and bounded in $\mathbb{R}^n$ with a smooth boundary $\partial\Omega$.

With a slight abuse of notation, we shall sometimes insert the variables x and t in the equations in order to make precise the dependence of various functions on these variables.

1.2.2 Zone controls in diffusion systems

We shall study the system (S_z) (where the subscript z stands for zone) described by the parabolic system

$$(S_z) \begin{cases} \dfrac{\partial y}{\partial t} - \Delta y = \sum\limits_{i=1}^{p} g_i(x) u_i(t) & \text{in } \Omega \times]0, T[\\ y(x, 0) = 0 & \text{in } \Omega \\ y(\xi, t) = 0 & \text{in } \partial\Omega \times]0, T[\end{cases}$$

(S_z) models a system excited by p zone controls $(\Omega_i, g_i)_{1 \leqslant i \leqslant p}$ with $g_i \in L^2(\Omega_i)$. Since the most realistic case is when the supports of the controls are disjoint, we assume that

$$\Omega_i \cap \Omega_j = \varnothing \qquad \text{for } i \neq j$$

The system (S_z) is a particular case of the system (S), where

$$Az = \Delta z \qquad z \in D(A) = H_0^1(\Omega) \cap H^2(\Omega)$$

and

$$B: [u_1, u_2, \ldots, u_p] \to \sum_{i=1}^{p} g_i u_i$$

B is a bounded linear map and A generates a strongly continuous semigroup $(S(t))_{t\geqslant 0}$ on $L^2(\Omega)$.

The system (S_z) may also be considered with Neumann or mixed boundary conditions. We shall examine some of these cases where there are important differences in the results.

Remark 1.7

We shall consider later (for $n = 1$) a system of the form (S_z) containing a transport term. The results on the location of the controls and controllability are not essentially changed.

1.2.3 Point controls in diffusion systems

We denote by (S_p) (the subscript p denotes point control) the system

$$(S_p)\begin{cases} \dfrac{\partial y}{\partial t} - \Delta y = \sum_{i=1}^{p} \delta(x - b_i)u_i(t) & \text{in } \Omega \times]0, T[\\ y(x, 0) = 0 & \text{in } \Omega \\ y(\xi, t) = 0 & \text{in } \partial\Omega \times]0, T[\end{cases}$$

(S_p) is a system excited by p point controls $(b_i, \delta_{b_i})_{1\leqslant i\leqslant p}$ localized at the points b_i in Ω. It is a particular case of system (S), with

$$Az = \Delta z \qquad z\in D(A) = H_0^1(\Omega)\cap H^2(\Omega)$$

and

$$B:\mathbb{R}^p \to D(A)^*$$

$$Bu = \sum_{i=1}^{p} \delta(x - b_i)u_i, \qquad u = [u_1, u_2, \ldots, u_p]$$

We shall develop results relative to this problem in Chapter 2, Section 2.2.

1.2.4 Boundary controls in diffusion systems

Two cases will be considered: boundary zone controls and point boundary controls.

Boundary zone controls

The system will be

$$(S_f)\begin{cases} \dfrac{\partial y}{\partial t} - \Delta y = 0 & \text{in } \Omega \times]0, T[\\ y(x, 0) = 0 & \text{in } \Omega \\ y(\xi, t) = \sum_{i=1}^{p} g_i(\xi)u_i(t) & \text{in } \partial\Omega \times]0, T[\end{cases}$$

(S_f) models a system excited by p control zones on the boundary $(\Gamma_i, g_i)_{1\leqslant i\leqslant p}$ with $\Gamma_i \subset \partial\Omega$ and $g_i\in L^2(\Gamma_i)$ for all i, $1 \leqslant i \leqslant p$. (S_f) has a unique weak solution given by

$$z(t) = -\sum_{i=1}^{p}\int_0^t AS(t-\tau)Gg_iu_i(\tau)\,\mathrm{d}\tau \tag{1.11}$$

where G is a Green's operator defined by

$$G: L^2(\partial\Omega) \to L^2(\Omega)$$
$$h \to Gh = y$$

$$\begin{array}{ll} \Delta y = 0 & \text{in } \Omega \\ y = h & \text{in } \partial\Omega \end{array}$$

A difficulty occurs in this case since, in general for $u \in L^2(0, T; \mathbb{R}^p)$, then $z(T) \notin L^2(\Omega)$. So, to have controllability in the sense of Definition 1.3 in $L^2(\Omega)$, it is possible to choose smoother controls (see Chapter 2, Section 2.3).

Point boundary controls

The system is described by

$$(S'_f) \begin{cases} \dfrac{\partial y}{\partial t} - \Delta y = 0 & \text{in } \Omega \times]0, T[\\ y(x, 0) = 0 & \text{in } \Omega \\ y(\xi, t) = \sum_{i=1}^{p} \delta(\xi - b_i) u_i(t) & \text{in } \partial\Omega \times]0, T[\end{cases}$$

(S'_f) represents a system excited on the boundary at p points b_i, $1 \leqslant i \leqslant p$. In order to define a solution of such a system, it is necessary to use the method of transposition proposed by Lions and Magenes (1968). Then (S'_f) has a unique weak solution such that $y(., T)$ makes sense in $H^{-\beta-1}(\Omega)$, $\beta > \frac{1}{2}$ when $u \in L^2(0, T; \mathbb{R}^p)$. We see that once again it is possible to choose smoother controls for $y(., T)$ to be in $L^2(\Omega)$.

The difficulty in the analysis of controllability questions rests essentially in the relation between the regularity of the controls and the state space $\mathscr{E}$. In practice, the states required are usually of the L^2 type. Thus, for systems of the type (S_f) and (S'_f), one can choose more regular controls.

1.3 STRATEGIC CONTROLS

The controllability properties of a system are affected by both the positioning of the controls and the distribution functions on their supports.

1.3.1 Definitions and Remarks

Definition 1.6

Suppose that $\mathscr{E}_1$ is a subspace of the state space $\mathscr{E}$. We say that the control (Ω_0, g_0) (or (b, δ_b)) is *strategic in* $\mathscr{E}_1$ if the system is exactly controllable to $\mathscr{E}_1$.

Definition 1.7

We say that the controls (Ω_0, g_0) (or (b, δ_b)) are *strategic* if the system is weakly controllable in $\mathscr{E}$.

Similar definitions apply to boundary controls. If the system is excited by p zone controls $(\Omega_i, g_i)_{1\leqslant i\leqslant p}$ (or point controls $(b_i, \delta_{b_i})_{1\leqslant i\leqslant p}$), then we say that the suite of controls $(\Omega_i, g_i)_{1\leqslant i\leqslant p}$ (or $(b_i, \delta_{b_i})_{1\leqslant i\leqslant p}$) is strategic if the system is weakly controllable in $\mathscr{E}$.

Remark 1.8

It is clear that, if the system is excited by p controls and for a certain i_0, $1\leqslant i_0\leqslant p$, the control with index i_0 is strategic, then so is the suite of p controls.

We shall now characterize strategic controls and as a result show that, in general, one control is not enough to ensure controllability. Then we go on to prove the existence of strategic controls for the case of zones (Ω_i, g_i), $1\leqslant i\leqslant p$. In particular, we show that, if the supports Ω_i are fixed, there always exist distributions g_i on Ω_i such that the control suite $(\Omega_i, g_i)_{1\leqslant i\leqslant p}$ is strategic.

In Chapter 2, we examine a number of examples where g_i are fixed and the locations of the support Ω_i are chosen so that the suite $(\Omega_i, g_i)_{1\leqslant i\leqslant p}$ is strategic.

1.3.2 Characterization of strategic controls

Consider the system

$$(S)\qquad \begin{cases} \dot{z} = Az + Bu \\ z(0) = 0 \end{cases}$$

Suppose that Hypothesis 1.3 holds and A has a complete set of orthonormal eigenfunctions (ϕ_{nj}) associated with the eigenvalues (λ_n), λ_n having multiplicity r_n. We then have the following characterization.

Proposition 1.2

The suite of controls $(\Omega_i, g_i)_{1\leqslant i\leqslant p}$ is strategic if and only if

(1) $p\geqslant \operatorname{Sup}_n(r_n)$;

(2) $\operatorname{rank}(G_n) = r_n$ for all n, where G_n is the matrix of order $p\times r_n$ with elements $(G_n)_{ij} = \langle g_i, \phi_{nj}\rangle_{L^2(\Omega_i)}$, $i=1,\dots,p$, $j=1,\dots,r_n$.

Proof We have seen in equation (1.3) that, for all $t\geqslant 0$ and all $z\in\mathscr{E}$,

$$S(t)z = \sum_n e^{\lambda_n t} \sum_{j=1}^{r_n} \langle z, \phi_{nj}\rangle \phi_{nj}$$

Hence the condition

$$\langle z, S(t)g_i\rangle_{L^2(\Omega_i)} = 0 \qquad \forall t\geqslant 0 \quad\text{and}\quad \forall i=1,\dots,p$$

implies that $z=0$ is equivalent to

$$\sum_n e^{\lambda_n t} \sum_{j=1}^{r_n} \langle g_i, \phi_{nj}\rangle\langle z, \phi_{nj}\rangle = 0 \qquad \forall t\geqslant 0 \quad\text{and}\quad \forall i=1,\dots,p$$

implies that $z=0$.

Necessity—suppose that the suite of controls is strategic. If for a certain n_0, $\operatorname{rank}(G_{n_0}) < r_{n_0}$, then

$$\exists v_{n_0} \in \mathbb{R}^{n_0}, \qquad v_{n_0} \neq 0 \qquad \text{such that } G_{n_0} v_{n_0} = 0$$

Now let

$$z_0 = \sum_{j=1}^{r_{n_0}} (v_{n_0})_j \phi_{n_0 j} \neq 0$$

Then

$$\langle z_0, S(t) g_i \rangle = 0 \qquad \forall t \geqslant 0 \quad \text{and} \quad i = 1, \ldots, p$$

Hence (S) is not controllable. So we have a contradiction.

Sufficiency—Suppose that, for a certain $z_0 \neq 0$,

$$\sum e^{\lambda_n t} \sum_{j=1}^{r_n} \langle z_0, \phi_{nj} \rangle \langle g_i, \phi_{nj} \rangle = 0$$

for all $t \geqslant 0$ and $i = 1, \ldots, p$.

Now for all $\lambda \in \mathbb{C}$ with $\operatorname{Re}(\lambda) > \omega_0$ for a certain ω_0, we have

$$\begin{aligned} 0 &= \int_0^\infty e^{-\lambda t} \langle z_0, S(t) g_i \rangle \, dt \\ &= \langle z_0, R(\lambda, A) g_i \rangle \end{aligned}$$

so that

$$\sum_n \frac{1}{\lambda - \lambda_n} \sum_{j=1}^{r_n} \langle z_0, \phi_{nj} \rangle \langle g_i, \phi_{nj} \rangle = 0 \qquad \forall i = 1, \ldots, p$$

and, by analyticity, the above holds for all $\lambda \neq \lambda_n$.

Consider in the complex plane

$$\Gamma_0 = \partial D(\lambda_{n_0}, \varepsilon)$$

(the boundary of the disc centred at λ_{n_0} with radius ε) where ε is sufficiently small that Γ_0 does not intersect another eigenvalue. By Cauchy's formula, we have

$$\begin{aligned} \frac{1}{2\pi i} \int_{\Gamma_0} \sum_n \frac{1}{\lambda - \lambda_n} \sum_{j=1}^{r_n} \langle z_0, \phi_{nj} \rangle \langle g_i, \phi_{nj} \rangle \, d\lambda &= \sum_{j=1}^{r_{n_0}} \langle z_0, \phi_{n_0 j} \rangle \langle g_i, \phi_{n_0 j} \rangle \\ &= 0 \qquad \forall i = 1, \ldots, p \end{aligned}$$

Now, for some $n_0 \in \mathbb{N}$, $\langle z_0, \phi_{n_0 j} \rangle \neq 0$. Hence, $\operatorname{rank}(G_{n_0}) < r_{n_0}$. ■

The above characterization shows, in particular, that the largest order of multiplicity of the eigenvalues must be finite.

1.3.3 Existence of Strategic Controls

Suppose that the system (S_z) is excited by p controls and the supports $(\Omega_i)_{1 \leqslant i \leqslant p}$ are fixed.

Proposition 1.3

Assume that $p \geqslant \mathrm{Sup}_n(r_n)$. For every open suite $(\Omega_i)_{1\leqslant i\leqslant p}$ in Ω, there exist functions $(g_i)_{1\leqslant i\leqslant p}$ such that

(1) $\mathrm{supp}(g_i) \subset \Omega_i$, $g_i \in L^2(\Omega_i)$ for all $i = 1, \ldots, p$;
(2) the suite of zone controls $(\Omega_i, g_i)_{1\leqslant i\leqslant p}$ is strategic.

Proof Let ϕ^i_{nj} be the restriction of ϕ_{nj} to Ω_i. Clearly, $\phi^i_{nj} \in L^2(\Omega_i)$. For fixed i, $1 \leqslant i \leqslant p$, $\{\phi^i_{nj}\}$ are dense in $L^2(\Omega_i)$. In effect, if $f_i \in L^2(\Omega_i)$ and

$$\tilde{f}_i = \begin{cases} f_i & \text{in } \Omega_i \\ 0 & \text{in } \Omega \backslash \Omega_i \end{cases}$$

if

$$\tilde{f}_i = \sum_n \sum_{j=1}^{r_n} \langle f_i, \phi_{nj} \rangle_{L^2(\Omega)} \phi_{nj}$$

Then by restriction to $\Omega_i(\phi_{nj}/\Omega_i = \phi^i_{nj})$ the series converges to f_i in $L^2(\Omega_i)$.

Since $p \geqslant \mathrm{Sup}_n(r_n)$, it is always possible to find a matrix $A = (a_{ij})$ of order $p \times r_n$ and rank $(A) = r_n$. Let $\tilde{A} = (\tilde{a}_{ij})$ where

$$\tilde{a}_{ij} = \frac{a_{ij}}{\sum_{i,j} a^2_{ij}}$$

Suppose that $\{f^i_{nj}\}$ is a complete orthonormal set in $L^2(\Omega_i)$ and consider the isomorphism $\theta_i: L^2(\Omega_i) \to L^2(\Omega_i)$ defined by

$$\theta_i(\phi^i_{nj}) = f^i_{nj}$$

for all $h \in L^2(\Omega_i)$, we have

$$\begin{aligned} \langle h, \phi^i_{nj} \rangle_{L^2(\Omega_i)} &= \langle h, \theta_i^{-1} f^i_{nj} \rangle_{L^2(\Omega_i)} \\ &= \langle \theta_i^{-1*} h, f^i_{nj} \rangle_{L^2(\Omega_i)} \end{aligned}$$

Now choose g_i such that

$$\theta_i^{-1*} g_i = \sum_n \sum_{j=1}^{r_n} \frac{\tilde{a}_{ij}}{\gamma_n} f^i_{nj} \qquad \gamma_n \neq 0, \quad \sum_n \frac{1}{\gamma_n^2} < \infty$$

then

$$\| g_i \|^2_{L^2(\Omega_i)} \leqslant \| \theta_i^* \|^2 \sum_n \sum_{j=1}^{r_n} \frac{|\tilde{a}_{ij}|^2}{\gamma_n^2} \leqslant K \sum_n \frac{1}{\gamma_n^2}$$

for some $K > 0$. Hence, $g_i \in L^2(\Omega_i)$ and by construction the suite of controls $(\Omega_i, g_i)_{1\leqslant i\leqslant p}$ is strategic (apply Proposition 1.2). ■

1.4 OBSERVABILITY AND SENSORS

We consider the system described by the equation

$$(S) \qquad \begin{cases} \dot{z} = Az + Bu & 0 < t < T \\ z(0) = 0 \end{cases}$$

where A generates a strongly continuous semigroup $(S(t))_{t\geqslant 0}$ and B is linear and bounded from the control space $\mathscr{U}$ to the state space $\mathscr{E}$.

1.4.1 Sensors

In general, it is not possible to measure all the state of a system during its evolution. The physical mechanism by which information about the state is obtained is called a 'sensor'. Just as in the case of controls, one may envisage different types of sensor.

If Ω is the geometric configuration on which the system (S) is modelled, then information about the state can be recovered via the following:

(1) zones $D_i \subset \Omega$ or $D_i \subset \partial\Omega$;
(2) points $x_i^m \in \Omega$ or $x_i^m \in \partial\Omega$;
(3) lines $L_i \subset \Omega$ or $L_i \subset \partial\Omega$.

(1) corresponds to a direct measurement over a part D_i of the domain Ω. (2) represents, for example, information obtained with a thermocouple, whereas (3) may represent, for example, measurements via a pivoting mirror. Perhaps the most important is that of point measurement.

In the case of zones, we shall suppose that the support D_i has a non-empty interior (Fig. 1.7).

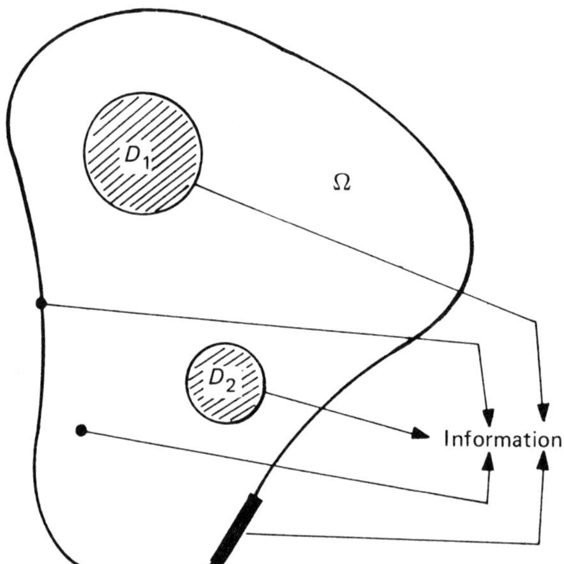

Fig. 1.7

Definition 1.8

Suppose that D_i is closed in Ω and $f_i \in L^2(D_i)$. A zone sensor is the pair (D_i, f_i) where D_i is the support of the sensor and f_i is the spatial distribution of the sensor over D_i.

Definition 1.9

Suppose that $x_i^m \in \Omega$. A point sensor is the pair $(x_i^m, \delta_{x_i^m})$ where $\delta_{x_i^m}$ is the Dirac mass at the point x_i^m.

For illustrations of point sensors in the form of thermocouples, see for example Amowoux *et al.* (1976), Amouroux and Babary (1982) and Pourciel *et al.* (1984).

Remark 1.9

In the case of sensors on the boundary $\partial\Omega$, the definitions are the same with $f_i \in L^2(\Gamma_i)$ where $\Gamma_i \subset \partial\Omega$ for zones or $x_i^m \in \partial\Omega$ for points. We say then boundary zone or boundary point sensors.

1.4.2 Output Function

Suppose that for the system (S) information in recovered via q sensors $(D_i, f_i)_{1 \leqslant i \leqslant q}$ in a time interval $[0, T_m]$. The output function is

$$(E) \qquad y(t) = Cz(t) \qquad 0 \leqslant t \leqslant T_m$$

We suppose that $y \in L^2(0, T_m; \mathbb{R}^q)$ and that C is linear from $\mathscr{E} \to \mathbb{R}^q$ (and, if it is not bounded, then $D(C) \subset \mathscr{E}$ and is invariant under the semigroup $(S(t))_{t \geqslant 0}$).

y is the vector function

$$y(t) = \begin{bmatrix} y_1(t) \\ \vdots \\ y_q(t) \end{bmatrix} = \begin{bmatrix} \langle f_1, z(t) \rangle_{L^2(D_1)} \\ \vdots \\ \langle f_q, z(t) \rangle_{L^2(D_q)} \end{bmatrix} \tag{1.12}$$

If A has a complete set of orthonormal eigenfunctions (ϕ_n), then

$$z(t) = \sum_n z_n(t)\phi_n$$

and so $y(t)$ is a q-dimensional vector with components of the form

$$y_i(t) = \sum_n z_n(t) \langle f_i, \phi_n \rangle_{L^2(D_i)} \qquad i = 1, \ldots, q$$

In the case of point sensors $(x_i^m, \delta_{x_i^m})$, then

$$y_i(t) = \sum_n z_n(t)\phi_n(x_i^m) \tag{1.13}$$

1.4.3 Observability

In a distributed physical system, it is not possible to measure the state of the system directly. The problem can be posed, therefore, of reconstruction of the state from the measurements—this is the problem of observability. The question is whether knowledge of the dynamics of the system, together with the inputs and outputs, is sufficient to reconstruct the initial state $z(0)$.

Since the systems which we consider are linear with output

$$y(t) = Cs(t)z(0) + C \int_0^t S(t-s)Bu(s)\,\mathrm{d}s$$

knowledge of the outputs and inputs yields the function

$$y(t) - C\int_0^t S(t-s)Bu(s)\,\mathrm{d}s$$

So we might as well consider the system to be autonomous.

Knowledge of the initial state is usually necessary to construct control algorithms. Normally, one consider a free system during a certain interval of time $[0, T_m]$ where measurements are taken via the sensors. Then these measurements are used to reconstruct the state. So the time interval is partitioned as follows.

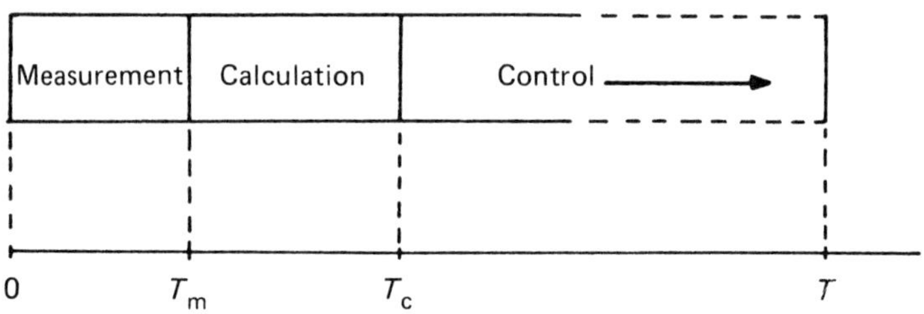

Fig. 1.8 Partition of the time interval

This partition is the subject of the following comments. The length of the interval $[T_m, T_c]$ is a function of the volume of calculations and the particular computers available to carry them out. So, in general, for a given system we may take $T_c - T_m$ as fixed, but then how should the interval $[T_m, T_c]$ be located in $[0, T]$. Many simulations for different systems have shown that the precision of the estimate of the initial state $z(0)$ increases with T_m and that the energy spent during the control interval $[T_c, T]$ is a decreasing function of $T - T_c$.

Suppose now that we have q sensors which furnish information on the system (S) during the interval $[0, T_m]$. We have

$$y(t) = CS(t)z(0) = K(t)z(0) \tag{1.14}$$

The observability problem is to determine an operator $G:\mathscr{Y} \to \mathscr{E}$ such that

$$z(0) = Gy$$

Equation (1.14) defines an operator K

$$K:\mathscr{E} \to L^2(0, T_m; \mathscr{O}) = \mathscr{Y}$$

K is linear and bounded and it is easy to see that

$$K^*:L^2(0, T_m; \mathscr{O}') \to \mathscr{E}'$$

with

$$K^*z = \int_0^{T_m} S^*(t)C^*z(t)\,\mathrm{d}t \tag{1.15}$$

As in controllability, because the system is infinite dimensional, there are many different notions of observability.

Definition 1.10 Exact observability

The system (*S*) augmented with the output equation (*E*) is said to be exactly observable on $[0, T_m]$ if

$$T_m(K^*) \supset \mathscr{E}' \tag{1.16}$$

Definition 1.11 Weak observability

The system (*S*) augmented with the output equation (*E*) is said to be weakly observable on $[0, T_m]$ if

$$\text{Ker}(K) = \{0\} \tag{1.17}$$

Remark 1.10

(1) In a similar manner, one says that we have exact observability in a space $\mathscr{E}_1 \supset \mathscr{E}$ if $I_m(K^*) \supset \mathscr{E}'_1$.
(2) In the following, instead of saying that the system (*S*) is observable, we say that the pair (A, C) is observable.

Now we give a characterization of the above properties in a similar way to that in Section 1.1 for controllability.

Property 1.3 Characterization of exact observability

The system (*S*) augmented with the output equation (*E*) is exactly observable on $[0, T_m]$ if and only if

$$\exists \gamma > 0 \qquad \text{such that } \| K z_0 \|_{L^2(0, T_m; \mathbb{R}^q)} \geqslant \gamma \| z_0 \|_{\mathscr{E}} \tag{1.18}$$

for all $z_0 \in \mathscr{E}$.

Proof This is identical with the proof of Property 1.1 where one replaces $\mathscr{E}'$ by $\mathscr{E}$ and $L^2(0, T; \mathbb{R}^m)$ by $L^2(0, T_m; \mathbb{R}^q)$. ■

Just as for exact controllability to a subspace $\mathscr{E}_1$, we can extend the above characterization to the case $\mathscr{E}_1 \supset \mathscr{E}$ by taking the norm in $\mathscr{E}_1$ in equation (1.18).

Property 1.4

The following statements are equivalent.

(1) (A, C) is weakly observable on $[0, T_m]$.
(2) $\overline{\text{Im}(K^*)} = \overline{\text{Im}(K^*K)} = (\mathscr{E})'$.

In effect, we have a characterization similar to that of Property 1.2.

Proof This is immediate since $\overline{\text{Im}(K^*)} = \text{Ker}(K^{\perp})$. ■

Suppose now that $\mathscr{E}$ is a Hilbert space and is identified with $\mathscr{E}'$ and consider the self-adjoint operator

$$N = K^*K : \mathscr{E} \to \mathscr{E}'$$

We have the following.

Property 1.5 Reconstruction of the initial stage

(A, C) is exactly observable on $[0, T_m]$ if and only if the operator N is positive definite and is weakly observable if and only if N is positive.

Proof

(1) If $N = K^*K$ is positive definite on a Hilbert space $\mathscr{E}$, there exists $k > 0$ such that

$$\langle Nz, z \rangle \geqslant k \| z \|^2 \qquad \forall z \in \mathscr{E}$$

$$\Leftrightarrow \quad \int_0^{T_m} \langle S^*(t)C^*CS(t)z \, dt, z \rangle \geqslant k \| z \|^2$$

$$\Leftrightarrow \int_0^{T_m} \| CS(t)z \|^2 \, dt \geqslant k \| z \|^2$$

$$\Leftrightarrow \quad \| Kz \|^2_{L^2(0, T_m; \mathbb{R}^q)} \geqslant k \| z \|^2$$

that is to say, the system is exactly observable on $[0, T_m]$.

(2) In the case where N is positive,

$$\langle Nz, z \rangle \geqslant 0 \qquad \forall z \in \mathscr{E} \quad \text{and} \quad \langle Nz, z \rangle = 0 \Rightarrow z = 0$$

If $Kz_0 = 0$, then

$$\int_0^{T_m} \| K(t)z_0 \|^2 \, dt = \int_0^{T_m} \langle CS(t)z_0, CS(t)z_0 \rangle \, dt = 0$$

and so $\langle Nz_0, z_0 \rangle = 0$ and then $z_0 = 0$. This proves the equivalence with weak observability on $[0, T_m]$. ■

In both (exact and weak) observability cases the operator N^{-1} exists and the initial state may be determined by the equation

$$z(0) = (K^*K)^{-1}K^*y \tag{1.19}$$

or formally

$$z(0) = \left\{ \int_0^{T_m} S^*(t)C^*CS(t) \, dt \right\}^{-1} \int_0^{T_m} S^*(t)C^*y(t) \, dt$$

The reconstruction operator G is

$$G = (K^*K)^{-1}K^* = K^\dagger \tag{1.20}$$

where $K^\dagger$ is the pseudo-inverse of K (see for example Chapter 3, section 3.1).

We have the following result.

Proposition 1.4

If the system (*S*) augmented with the output equation (*E*) is exactly observable on $[0, T_m]$, then the application

$$G: L^2(0, T_m; \mathbb{R}^q) \to \mathscr{E}$$
$$y \to z(0)$$

is continuous.

Proof (A, C) is exactly observable on $[0, T_m]$ if and only if there exists $k > 0$ such that

$$\langle K^*Kz \rangle \geqslant k \| z \|^2 \qquad \forall z \in \mathscr{E}$$

However, a bounded linear operator T has a continuous linear inverse if and only if there exists $m > 0$ such that

$$\| Tz \| \geqslant m \| z \| \qquad \forall z \in \mathscr{E}$$

Now

$$\| K^*Kz \| \, \| z \| \geqslant \langle K^*Kz, z \rangle \geqslant k \| z \|^2$$

Hence

$$\| K^*Kz \| \geqslant k \| z \|$$

and this proves the result. ■

Proposition 1.4 shows that, if a system is exactly observable, then the construction of the initial state can be carried out in a satisfactory manner, since it varies continuously with the measurement. A small error in the measurement will result in only small errors in the initial state.

In the case where the system is weakly observable, this dependence is not continuous, which means the application of equation (1.19) may result in large errors in the initial state even though there may be only small measurement errors. Nevertheless, it is possible in this case to approximate the operator $(K^*K)^{-1}$ by a continuous operator. In effect, equation (1.19) can be obtained via the minimization problem with criterion

$$J(z_0) = \int_0^{T_m} \| y(t) - CS(t)z_0 \|^2 \, dt \tag{1.21}$$

This represents the error between the measurement and the output for each initial state z_0. Then,

$$\min \{J(z_0)\} = J(z_0^*), \qquad z_0^* = (K^*K)^{-1} K^*y \tag{1.22}$$

However, z_0^* may not lie in $\mathscr{E}$. We regularize the problem by introducing a criterion

$$J_\varepsilon(z_0) = J(z_0) + \varepsilon \langle Rz_0, z_0 \rangle \tag{1.23}$$

where $\varepsilon > 0$ and R is self-adjoint and positive definite. The minimization of equation (1.23) yields

$$z_0^\varepsilon = (K^*K + \varepsilon R)^{-1} K^*y \tag{1.24}$$

It is easy to see that the operator

$$G_\varepsilon = (K^*K + \varepsilon R)^{-1} K$$

is bounded from $\mathscr{Y}$ to $\mathscr{E}$ and hence $z_0^\varepsilon \in \mathscr{E}$ and varies continuously with $y \in \mathscr{Y}$. Then, if the system is weakly observable, one can show that

$$\lim_{\varepsilon\to 0}(z_0^\varepsilon) = z(0) \tag{1.25}$$

provided that $z(0)\in\mathscr{E}$. For more details, see Lions (1968) and Kobayashi (1976).

1.4.4 Strategic sensors

Just as for controllability, we shall now show that the observability properties of a system are affected by the choice of sensors (number, location and spatial distribution).

We always consider the system (S):

$$(S)\qquad \begin{cases} \dot{z} = Az + Bu \\ z(0) = z_0 \end{cases}$$

with output equation

$$(E)\qquad y = Cz$$

Ω is the geometric domain on which S is modelled. D is a closed set in Ω or its boundary. $f\in L^2(D)$. $[0, T_{\rm m}]$ is the interval of measurement. We suppose that the measurement is made via a single sensor (D, f); then

$$y(t) = \langle f, z(t)\rangle_{L^2(D)}$$

If (ϕ_n) is a complete orthonormal set of eigenfunctions for A, then

$$y(t) = \sum_n z_n(t)\langle f, \phi_n\rangle_{L^2(D)}$$

Definition 1.12 Strategic sensor

We say that the sensor (D, f) is strategic if the system (S) augmented with the output equation (E) is weakly observable on $[0, T_{\rm m}]$.

Remark 1.11

(1) The definition is the same for a point sensor with D reduced to the point x^m and $f = \delta_{x^m}$.
(2) In the case of q sensors, we say that the suite of sensors $(D_i, f_i)_{1\leqslant i\leqslant q}$ is strategic if the system (S) augmented with the output equation (E) is weakly observable on $[0, T_{\rm m}]$.

For distributed parameter systems the notion of exact observability is, in general, too strong. This is why we have adopted the notion of weak observability in the above definition of strategic sensors.

Example 1.2

Suppose that $\Omega =]0, 1[$ and consider the system described by the equations

$$\begin{aligned} &\frac{\partial z}{\partial t} = \alpha\frac{\partial^2 z}{\partial x^2} + \gamma z \qquad \text{in } \Omega\times]0, T[\\ &z(0, t) = z(1, t) = 0 \end{aligned} \tag{1.26}$$

The measurements are taken to be q point sensors $(x_i^m, \delta_{x_i^m})_{1 \leq i \leq q}$ in Ω, so that the output equation is

$$y(t) = \begin{bmatrix} y(x_1^m, t) \\ \vdots \\ y(x_q^m, t) \end{bmatrix} \tag{1.27}$$

We have the following result.

Proposition 1.5

The suite of sensors $(x_i^m, \delta_{x_i^m})_{1 \leq i \leq q}$ is strategic for the system (1.26) with output equation (1.27) if x_i^m is irrational for one i_0, $1 \leq i_0 \leq q$.

Proof For the operator

$$\alpha \frac{\partial^2}{\partial x^2} + \gamma$$

with Dirichlet boundary conditions, the eigenfunctions are

$$\phi_n(x) = 2 \sin(n\pi x)$$

with corresponding eigenvalues

$$\lambda_n = \gamma - \alpha n^2 \pi^2$$

The output function (1.27) has the ith coordinate

$$y(x_i^m, t) = \sum_n z_n(t)\phi_n(x_i^m)$$

where

$$\begin{aligned} z_n(t) &= \langle z(.\,, t), \phi_n(.) \rangle_{L^2(\Omega)} \\ &= \langle z(0), \phi_n \rangle e^{\lambda_n t} \end{aligned}$$

Now suppose that $x_{i_0}^m$ is irrational; then $\phi_n(x_{i_0}^m) \neq 0$ for all $n = 1, 2, \ldots$. So, just as in the proof of Proposition 1.2,

$$\sum_n \langle z(0), \phi_n \rangle e^{\lambda_n t} \phi_n(x_{i_0}^m) = 0 \qquad \forall t \in [0, T_m]$$

implies that $\langle z(0), \phi_n \rangle = 0$ for all $n \in \mathbb{N}$. Thus the system is weakly observable. ■

Remark 1.12

We have an analogous result to Proposition 1.5 in the case where the sensors are of the zone type (D_i, f_i) with one of the centres of the interval D_i being irrational and the corresponding f_i being symmetric about x_i^m. For other situations, see Chapter 2.

1.5 STABILIZABILITY AND DETECTABILITY

1.5.1 Stabilizability

One of the most important considerations in the analysis of control systems is that of stability. Consider again the system (S):

$$(S)\qquad \begin{cases} \dot{z} = Az + Bu & 0 < t < T \\ z(0) = z_0 \end{cases}$$

where A generates a strongly continuous semigroup $(S(t))_{t \geqslant 0}$ on $\mathscr{E}$.

Definition 1.13 Exponentially stable semigroup

The semigroup $(S(t))_{t \geqslant 0}$ is said to be exponentially stable if there exist two positive constants M and ω such that

$$\| S(t) \| \leqslant M \mathrm{e}^{-\omega t} \qquad t \geqslant 0 \tag{1.28}$$

Note that, if $(S(t))_{t \geqslant 0}$ is an exponentially stable semigroup, then for all $z_0 \in \mathscr{E}$, the solution of the autonomous system

$$\begin{aligned} \dot{z} &= Az \\ z(0) &= z_0 \end{aligned} \tag{1.29}$$

satisfies

$$\| z(t) \| = \| S(t) z_0 \| \leqslant M \mathrm{e}^{-\omega t} \| z_0 \|$$

Example 1.3

Suppose that (ϕ_{nj}) is a complete orthonormal system of eigenfunctions of A with associated eigenvalues λ_n each of multiplicity r_n. We have already seen that

$$S(t)z = \sum_n \mathrm{e}^{\lambda_n t} \sum_{j=1}^{r_n} \langle z, \phi_{nj} \rangle \phi_{nj}$$

and so

$$\| S(t)z \|^2 = \sum_n \mathrm{e}^{2\mathrm{Re}(\lambda_n t)} \sum_{j=1}^{r_n} \langle z, \phi_{nj} \rangle^2 \tag{1.30}$$

Suppose that there exist $\varepsilon > 0$ such that

$$\mathrm{Re}(\lambda_n) < -\varepsilon \qquad \forall n \geqslant 1$$

then

$$\begin{aligned} \| S(t)z \|^2 &\leqslant \mathrm{e}^{-2\varepsilon t} \sum_n \sum_{j=1}^{r_n} \langle z, \phi_{nj} \rangle^2 \\ &= \mathrm{e}^{-2\varepsilon t} \| z \|^2 \end{aligned}$$

Hence

$$\| S(t) \| \leqslant \mathrm{e}^{-\varepsilon t}$$

and so the semigroup is exponentially stable.

Suppose that, for some i,

$$\mathrm{Re}(\lambda_i) > 0$$

then the semigroup is not exponentially stable. In effect, it suffices to consider $z_0 = \phi_{i1}$; then

$$z(t) = S(t) z_0 = \mathrm{e}^{\lambda_i t} \phi_{i1}$$

Remark 1.13

Example 1.3 suggests that the stability of a system is associated with the spectrum

$\sigma(A)$ of the operator A. In fact, one can show that

$$\operatorname{Sup}\{\operatorname{Re}(\lambda):\lambda\in\sigma(A)\}\leqslant\inf_{\omega}\{\omega:\|S(t)\|\leqslant M\mathrm{e}^{\omega t}\} \tag{1.31}$$

Thus, if the semigroup is exponentially stable, then necessarily the spectrum of A lies in the half-plane

$$\operatorname{Re}(\lambda)<0$$

Generally, we do not have equality in equation (1.31) but, for most real systems, and all these we consider here, the equality holds. As a consequence the exponential stability of a semigroup may be determined by examining $\sigma(A)$.

In the case where A does not generate an exponentially stable semigroup, it may be possible to remedy this by suitable control action.

Definition 1.14 Stabilizability

The system (S) is said to be stabilizable (or the pair (A, B) is said to be stabilizable) if there exists a feedback control

$$u=-Fz \tag{1.32}$$

such that $A-BF$ generates a strongly continuous semigroup $(S_F(t))_{t\geqslant 0}$ which is exponentially stable.

Remark 1.14

The control defined in equation (1.32) is state feedback and, in order to implement it, the whole state must be known at each instant of time $t>0$. This is rarely possible and so it is necessary to estimate the state via the sensors. Moreover, this must be carried out continually since the system will, in general, be constantly subjected to perturbations. We return to this question later.

In finite dimensional linear theory (Kwakernaak and Sivan, 1972; Kailath, 1980; Hinrichsen and Pritchard, 1986), there is a certain equivalence between controllability and stabilizability. For infinite dimensional systems the situation is much more complex. However, it is possible to show that, if at some time T it is possible to steer the system exactly to the null state with a control $u\in L^2(0, T; \mathscr{U})$, then it is possible to stabilize the system.

Recall that the null state may be reached if

$$\operatorname{Im}(H)\supset\{0\} \tag{1.33}$$

where $H: L^2(0, T; \mathscr{U})\to\mathscr{E}$ is defined in equation (1.2) by

$$Hu=\int_0^T S(T-t)Bu(t)\,\mathrm{d}t$$

Equation (1.33) is equivalent to (Property 1.1)

$$\exists\gamma>0 \qquad \text{such that } \|B^*S^*(.)z^*\|_{L^2(0,T;\mathscr{U}')}\geqslant\gamma\|S^*(T)z^*\|_{\mathscr{E}'} \tag{1.34}$$

for all $z^*\in\mathscr{E}'$.

For parabolic systems the condition (1.34) is much less restrictive than that of exact controllability given by equation (1.9).

We shall now characterize the notion of stabilizability via the structure of the controls.

1.5.2 Stabilizability and controls

We shall always assume that (ϕ_{nj}) are a complete set of orthonormal eigenfunctions of A with corresponding eigenvalues λ_n of multiplicity r_n. Then, we are able to characterize stabilizability via the choice of controls in a similar manner to that of Proposition 1.2 for controllability. The main difference is that we can restrict our considerations to those parts of the spectrum which are positive real.

Proposition 1.6

Suppose that the system is excited by p zone controls $(\Omega_i, g_i)_{1 \leqslant i \leqslant p}$ and $\sigma(A)$ has J eigenvalues with non-negative real parts. Then the system (S) is stabilizable if and only if

(1) $p \geqslant \sup_{1 \leqslant n \leqslant J}(r_n)$;
(2) $\operatorname{rank}(G_n) = r_n$ for all $n = 1, 2, \ldots, J$ where $(G_n)_{ij} = \langle g_i, \phi_{nj} \rangle_{L^2(\Omega_i)}$, $i = 1, \ldots, p$, $j = i, \ldots, r_n$.

Proof Projecting the system (S) on its unstable and stable parts $z = [z_{\mathrm{u}}, z_{\mathrm{s}}]$ gives

$$\dot{z}_{\mathrm{u}} = A_{\mathrm{u}} z_{\mathrm{u}} + PBu$$
$$\dot{z}_{\mathrm{s}} = A_{\mathrm{s}} z_{\mathrm{s}} + (I - P)Bu$$

A_{u} is represented by a matrix of order

$$\left(\sum_{n=1}^{J} r_n, \sum_{n=1}^{J} r_n \right)$$

given by

$$A_{\mathrm{u}} = \operatorname{diag}(\lambda_1, \ldots, \lambda_1, \lambda_2, \ldots, \lambda_J, \ldots, \lambda_J)$$

and

$$PB = [G_1^{\mathrm{T}} G_2^{\mathrm{T}} \cdots G_J^{\mathrm{T}}]^{\mathrm{T}}$$

Under condition (2), it is easy to see that the finite dimensional system is controllable and hence stabilizable. So there exists a control $u = -Fz_{\mathrm{u}}$ such that

$$\| \mathrm{e}^{(A_{\mathrm{u}} - PBF)t} \| \leqslant M\mathrm{e}^{-\delta t} \qquad M, \delta > 0$$

Hence

$$\| z_{\mathrm{u}}(t) \| \leqslant M\mathrm{e}^{-\delta t} \| Pz_0 \|$$

and

$$\| u(t) \| \leqslant M\mathrm{e}^{-\delta t} \| F \| \, \| Pz_0 \|$$

Now the semigroup generated by A_{s} is exponentially stable, and so there exist $\bar{M}$,

$\bar{\delta} > 0$ such that

$$\begin{aligned}\| z_s(t) \| &\leq \bar{M} e^{-\delta t} \| (I-P) z_0 \| \\ &\quad + \int_0^t \bar{M} e^{-\bar{\delta}(t-\tau)} \| (I-P) B \| \, \| u(\tau) \| \, d\tau \\ &\leq \bar{M} e^{-\bar{\delta} t} \| (I-P) z_0 \| \\ &\quad + \bar{M} M \| F \| \, \| (I-P) B \| \, \| P z_0 \| \int_0^t e^{-\bar{\delta}(t-\tau)} e^{-\delta t} \, d\tau\end{aligned}$$

This completes the proof. ■

Applications of this result will be presented in the next chapter.

1.5.3 Detectability

Detectability is in some sense a dual notion to stabilizability. Consider the system

$$(S) \qquad \begin{cases} \dot{z} = Az + Bu \\ z(0) = z_0 \end{cases}$$

augmented with the output equation

$$(E) \qquad y = Cz$$

Again we assume that A generates a strongly continuous semigroup $(S(t))_{t \geq 0}$.

Definition 1.15 Detectability

The system (S) augmented with the output equation (E) is said to be detectable (or the pair (A, C) is detectable) if there exists an operator

$$K : \mathcal{O} \to \mathscr{E}$$

such that $A - KC$ generates a strongly continuous semigroup $(S_K(t))_{t \geq 0}$ which is exponentially stable.

It is not immediately clear why the notion of detectability is important for a system. The motivation is that, for a detectable system, it is possible to construct an asymptotic estimate of the state of the system. Consider the system

$$\begin{aligned}\dot{w} &= Aw + Bu + K(y - Cw) \\ w(0) &= 0\end{aligned} \tag{1.35}$$

Then the error $e = z - w$ satisfies

$$\begin{aligned}\dot{e} &= (A - KC)e \\ e(0) &= z_0\end{aligned}$$

so that, if the system is detectable, it is possible to choose K to achieve

$$\| e(t) \| = \| S_K(t) e_0 \| \to 0 \qquad \text{as } t \to \infty$$

Clearly the observer (1.35) is infinite dimensional. It is possible to show that a system is detectable if it is exactly *finally* observable in $[0, T]$. By this we mean that

$$\exists \gamma > 0 \qquad \text{such that } \| CS(.)z \|_{L^2(0,T;\mathcal{O})} \geqslant \gamma \| S(T)z \|_{\mathscr{E}} \tag{1.36}$$

for all $z \in \mathscr{E}$. For parabolic systems, this notion is far less restrictive than that of exact observability characterized by equation (1.18).

1.5.4 Detectability and sensors

Using duality, we have the following characterization of detectability in terms of the structure of the sensors.

Proposition 1.7

Suppose that information in retrieved from the system via q zone sensors $(D_i, f_i)_{1 \leqslant i \leqslant q}$ and the spectrum of A contains J eigenvalues with non-negative real parts. Then (A, C) is detectable if and only if

(1) $q \geqslant \sup\limits_{1 \leqslant n \leqslant J} (r_n)$;

(2) $\operatorname{rank}(G_n) = r_n$ for $n = 1, 2, \ldots, J$
where $(G_n)_{ij} = \langle f_i, \phi_{nj} \rangle_{L^2(D_i)}$, $i = 1, \ldots, q$, $j = 1, \ldots, r_n$. ■

In both Proposition 1.6 and Proposition 1.7 the main difference with the characterizations of strategic controls and sensors is that we are able to restrict to the unstable part of the system.

1.6 COMPENSATORS

1.6.1 Generalities

In the preceding section, we considered the problem of stabilization via state feedback. We shall now consider the possibility of stabilization via the output of the system. We shall not consider direct feedback of the output but instead pass it through a dynamic system of the form

$$\begin{aligned} \dot{w} &= \bar{A}w + Ky + f \\ w(0) &= 0 \end{aligned} \tag{1.37}$$

Then we generate the control via

$$u = -Fw \tag{1.38}$$

The above scheme consisting of equations (1.37) and (1.38) is called a compensator

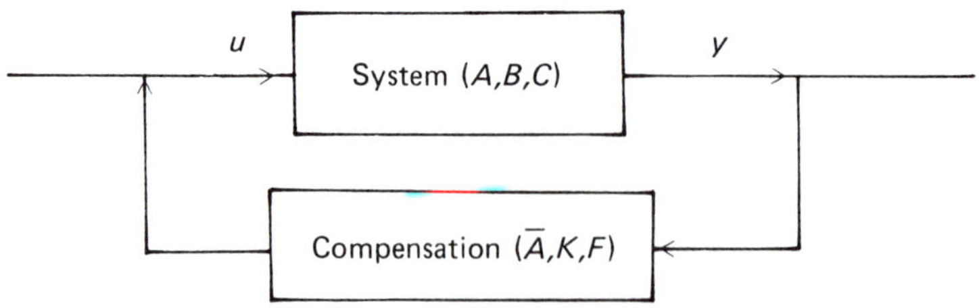

Fig. 1.9 Compensator system

(Fig. 1.9) and may be implemented in a closed-loop form by means of a numerical simulation of equations (1.37).

The purpose of the compensator is to ensure stability, for which we have the following result.

Proposition 1.8

If (A, B) is stabilizable and (A, C) is detectable, then the compensator

$$\begin{aligned} \dot{w} &= (A - KC)w + Bu + KCy \\ u &= -Fw \end{aligned} \tag{1.39}$$

ensures stability of the system (S).

Proof Let $e = z - w$; then

$$\dot{e} = (A - KC)e$$
$$\dot{z} = (A - BF)z + BFe$$

Since (A, B) is stabilizable and (A, C) is detectable, there exist two operators F and K such that $A - BF$ and $A - KC$ generate strongly continuous semigroups $(S_F(t))_{t\geqslant 0}$ $(S_K(t))_{t\geqslant 0}$ which are exponentially stable. Hence there exist M_K, $\omega_K > 0$, such that

$$\| S_K(t) \| \leqslant M_K \mathrm{e}^{-\omega_K t} \qquad t \geqslant 0$$

and $M_F, \omega_F > 0$, such that

$$\| S_F(t) \| \leqslant M_F \mathrm{e}^{-\omega_F t} \qquad t \geqslant 0$$

However,

$$z(t) = S_F(t)z_0 + \int_0^t S_F(t-s)BFS_K(s)e_0 \, \mathrm{d}s$$

Hence,

$$\| z(t) \| \leqslant M_F \mathrm{e}^{-\omega_F t} \| z_0 \| + M_F M_K \| BF \| \, \| e_0 \| \int_0^t \mathrm{e}^{-\omega_F (t-s)} \mathrm{e}^{-\omega_K s} \mathrm{d}s$$

Thus, by choosing $\omega_F \neq \omega_K$, $\| z(t) \|$ is bounded by a negative exponential function of t. ■

1.6.2 Finite dimensional compensators

The compensator (1.39) is infinite dimensional and so it cannot be implemented on a computer. The question arises, therefore, whether or not it is possible to construct a finite dimensional compensator which stabilizes the system.

In order to develop this, let us suppose that the spectrum $\sigma(A) = \{\lambda_n\}$ is decomposed into the form

$$\begin{aligned} \sigma(A) &= \sigma_{\mathrm{u}}(A) \cup \sigma_{\mathrm{s}}(A) \\ \sigma_{\mathrm{u}}(A) &= \sigma(A) \cap \{\lambda \in \mathbb{C} : \mathrm{Re}(\lambda) \geqslant -\delta, \, \delta > 0\} \\ &= \{\lambda_1, \lambda_2, \ldots, \lambda_J\} \\ \sigma_{\mathrm{s}}(A) &= \sigma(A) \cap \{\lambda \in \mathbb{C} : \mathrm{Re}(\lambda) < -\delta\} \\ &= \{\lambda_{J+1}, \ldots\} \end{aligned}$$

The corresponding decomposition of the system has the form

$$z=\begin{bmatrix} z_u \\ z_s \end{bmatrix}, \qquad A=\begin{bmatrix} A_u & 0 \\ 0 & A_s \end{bmatrix}, \qquad B=\begin{bmatrix} B_u \\ B_s \end{bmatrix}$$

and

$$(S) \qquad \begin{cases} \dot{z}_u = A_u z_u + B_u u \\ \dot{z}_s = A_s z_s + B_s u \end{cases}$$

Similarly, we have

$$(E) \qquad y = C_u z_u + C_s z_s = [C_u\, C_s]\begin{bmatrix} z_u \\ z_s \end{bmatrix}$$

We shall give a condition for the existence of a finite dimensional compensator which stabilizes (S).

Proposition 1.9

Suppose that

(1) (A_u, B_u) is stabilizable: there exists F_u given by

$$F_u : \mathbb{R}^{\sum_{i=1}^{J} r_i} \to \mathbb{R}^p$$

such that $A_u - B_u F_u$ is stable.

(2) (A_u, C_u) is detectable: there exists K_u given by

$$K_u : \mathbb{R}^q \to \mathbb{R}^{\sum_{i=1}^{J} r_i}$$

such that $A_u - K_u C_u$ is stable.

Then there exist $\bar{M}$, $\bar{\alpha} > 0$ such that

$$\bar{A} = \begin{bmatrix} A_u - K_u C_u & 0 & 0 \\ B_u F_u & A_u - B_u F_u & 0 \\ B_s F_u & -B_s F_u & A_s \end{bmatrix}$$

generates a strongly continuous semigroup $(\bar{S}(t))_{t \geqslant 0}$ with

$$\| \bar{S}(t) \| \leqslant \bar{M}\,\mathrm{e}^{-\bar{\alpha} t}$$

If J, K_u and C_s are such that

$$\sup_{w \in \mathbb{R}} \{ \| C_s(\mathrm{i}w - A_s)^{-1} B_s F_u (\mathrm{i}w - A_u + B_u F_u)^{-1} (\mathrm{i}w - A_u)(\mathrm{i}w - A_u + K_u C_u)^{-1} K_u \| \} < 1 \tag{1.40}$$

then there exists a finite dimensional compensator of dimension $\sum_{i=1}^{J} r_i$ which stabilizes (S),

Proof The first part of the theorem is proved using perturbation techniques (Schumacher, 1981). For the second part, we have if

$$\begin{aligned} \dot{w} &= (A_u - K_u C_u) w + K_u y - B_u F_u w \\ u &= -F_u w \end{aligned} \tag{1.41}$$

then

$$\begin{bmatrix} \dot{e} \\ \dot{z}_u \\ \dot{z}_s \end{bmatrix} = \begin{bmatrix} A_u - K_u C_u & 0 & K_u C_s \\ B_u F_u & A_u - B_u F_u & 0 \\ B_s F_u & -B_s F_u & A_s \end{bmatrix} \begin{bmatrix} e \\ z_u \\ z_s \end{bmatrix}$$

where $e = z_u - w$. We write the above equation in the form

$$\begin{bmatrix} \dot{e} \\ \dot{z}_u \\ \dot{z}_s \end{bmatrix} = \bar{A} \begin{bmatrix} e \\ z_u \\ z_s \end{bmatrix} + \begin{bmatrix} K_u \\ 0 \\ 0 \end{bmatrix} [0 \quad 0 \quad C_s] \begin{bmatrix} e \\ z_u \\ z_s \end{bmatrix}$$

Then using a result of Hinrichsen and Pritchard (1986) the perturbed system will be exponentially stable provided that

$$\sup_{w \in \mathbb{R}} \left\{ \left\| [0 \quad 0 \quad C_s](iw - \bar{A})^{-1} \begin{bmatrix} K_u \\ 0 \\ 0 \end{bmatrix} \right\| \right\} < 1$$

The result follows by calculating the inverse of $iw - \bar{A}$, ■

1.6.3 Remarks

Remark 1.15

If the output and input operators C and B are bounded, it is possible to make $\| C_s \|$ and $\| B_s \|$ small by choosing J large.

Example 1.4

Suppose that the system (S) is excited by one control (Ω_1, g_1) and there is one sensor (D_1, f_1). For simplicity, assume that $r_i = 1$, and $\lambda_1 > 0$, $0 > \lambda_2 > \lambda_3, \ldots$. We set

$$b_i = \langle g_1, \phi_i \rangle_{L^2(\Omega_1)}, \qquad c_i = \langle f_1, \phi_i \rangle_{L^2(D_1)}$$

Suppose that f_u and k_u are chosen so that

$$\lambda_1 - b_1 f_u = -\lambda_u, \qquad \lambda_1 - k_u c_1 = -\lambda_c$$

where $\lambda_1 > \lambda_u, \lambda_c > 0$. Then, equation (1.40) will certainly be satisfied if

$$\sup_{w \in \mathbb{R}} \left\{ \left| \sum_{i=2}^{\infty} c_i (iw - \lambda_i)^{-1} b_1 \right| \right\} < \frac{\lambda_u \lambda_c}{(\lambda_1 + \lambda_u)(\lambda_1 + \lambda_c)}$$

■

Remark 1.16

If the system is stabilizable and detectable, it is possible to construct a compensator but, in order to have a finite dimensional compensator, it is necessary to ensure conditions (1.40). This condition will certainly hold if

$$\sup_{w \in \mathbb{R}} \{ \| C_s (iw - A_s)^{-1} B_s \| \} < \inf_{w \in \mathbb{R}} [\| \{ I + G_B(iw) \}^{-1} F_u K_u \{ I + G_C(iw) \}^{-1} \|^{-1}]$$

where $G_B(\mathrm{i}w) = F_\mathrm{u}(\mathrm{i}w - A_\mathrm{u})^{-1}B_\mathrm{u}$ and $G_C = C_\mathrm{u}(\mathrm{i}w - A_\mathrm{u})^{-1}K_\mathrm{u}$. Clearly, this condition is satisfied if $C\phi_{nj} = 0$, $n = J+1,\ldots$, $j = 1,\ldots,r_n$, or $B^*\phi_{nj} = 0$, $n = J+1,\ldots$, $j = 1,\ldots,r_n$. However, if this is not possible by the choice of controls and sensors, then one should try to choose the parameters governing their structure (see Example 1.4) so that $\sup\{\|C_\mathrm{s}(\mathrm{i}w - A_\mathrm{s})^{-1}B_\mathrm{s}\|$ is as small as possible.

Chapter 2

Relation between various notions underlying the analysis of systems and the choice of controls and sensors

In this chapter, we consider parabolic systems on various domains and will be mainly concerned in conditions of symmetry. We develop many results linking the controls with controllability; then these results will be applied to the problems of sensors and observability and also to stabilizability and detectability. We shall be particularly interested in point controls and sensors, and these will be the subject of a separate section. We also consider an example where the domain Ω is not bounded.

2.1 ZONE CONTROLS

Consider the system (S_z)

$$(S_z) \qquad \begin{cases} \dfrac{\partial y}{\partial t} - \Delta y = \sum_{i=1}^{p} g_i u_i & \text{in } \Omega \times]0, T[\\ y(x, 0) = 0 & \text{in } \Omega \\ y = 0 & \text{in } \partial\Omega \times]0, T[\end{cases}$$

(S_z) is excited by p zone controls $(\Omega_i, g_i)_{1 \leqslant i \leqslant p}$ where $\Omega_i \subset \Omega$ is the support of the ith control and $g_i \in L^2(\Omega_i)$ defines the distribution of the control over Ω_i. Proposition 1.2 gives a characterization of a strategic suite, namely

(1) $p \geqslant \sup_n (r_n)$;

(2) $\operatorname{rank}(G_n) = r_n$ where $(G_n)_{ij} = \langle g_i, \phi_{nj} \rangle_{L^2(\Omega_i)}$, $1 \leqslant i \leqslant p$, $1 \leqslant j \leqslant r_n$,

ϕ_{nj} are the eigenfunctions of Δ with Dirichlet boundary conditions and r_n is the multiplicity of the eigenvalue λ_n.

This shows that the notion of controllability is linked with the spectrum of Δ. For the case where $r_n = 1$ for all n we have the following simple result.

Proposition 2.1

Suppose that the system (S_z) is excited by one control and $r_n = 1$ for all n. This control is strategic if and only if all the coordinates of g_i with respect to the eigenfunctions ϕ_n (i.e. $\langle g_1, \phi_n \rangle_{L^2(\Omega_1)}$) are non-zero.

Proof This is immediate since $\operatorname{rank}(G_n) = \operatorname{rank} \langle g_1, \phi_n \rangle_{L^2(\Omega_1)}$. ■

Remark 2.1

This result is also true for the system (S_z) with Neumann boundary conditions.

2.1.1 Zone controls strategic for one-dimensional domains

We consider the case where $\Omega =]0, a[$; then the eigenfunctions for the Dirichlet problem are

$$\phi_n(x) = \left(\frac{2}{a}\right)^{1/2} \sin\left(\frac{n\pi x}{a}\right) \qquad n \geqslant 1 \tag{2.1}$$

and the eigenvalues are

$$\lambda_n = -\frac{n^2\pi^2}{a^2} \qquad n \geqslant 1 \tag{2.2}$$

In this case the multiplicity $r_n = 1$ for all n and so we may envisage the possibility of controllability with one control.

Consider then the system

$$\begin{aligned} \frac{\partial y}{\partial t} &= \frac{\partial^2 y}{\partial x^2} + g_1 u_1 && \text{in }]0, a[\times]0, T[\\ y(x, 0) &= 0 && \text{in }]0, a[\\ y(0, t) &= y(a, t) = 0 && \text{in }]0, T[\end{aligned} \tag{2.3}$$

and $\operatorname{supp}(g_1) = [x_1 - l_1, x_1 + l_1] \subset]0, a[$, then we have the following.

Proposition 2.2

If g_1 is symmetric about the point x_1, then the control $([x_1 - l_1, x_1 + l_1], g_1)$ is not strategic when $x_1/a \in \mathbb{Q}$ (the rational numbers).

Proof We use Proposition 2.1 since $r_n = 1$ for all n. If x_1/a is rational, there exists a natural number n_0 such that $n_0 x_1/a \in \mathbb{N}$. However, then $\phi_{n_0}(x_1) = 0$ and, if g_1 is symmetric about x_1,

$$\langle g_1, \phi_{n_0} \rangle = \int_{x_1 - l_1}^{x_1 + l_1} g_1(x) \phi_{n_0}(x) \, \mathrm{d}x = 0$$

Hence, $\operatorname{rank}(G_{n_0}) = 0$. ■

Physically the hypothesis of symmetry is realistic. It is the case, for example, when the control is uniformly distributed over its domain ($g_1 = \alpha_1 \chi_{\Omega_1}$) such as in the case of a non-point heater (Souza Leao *et al.*, 1983; Babary and Souza Leao, 1984).

The proposition indicates bad positions of the control.

Remark 2.2

A similar result holds for systems of the form

$$\begin{aligned} \frac{\partial y}{\partial t} &= \alpha \frac{\partial^2 y}{\partial x^2} + \beta \frac{\partial y}{\partial x} + \gamma y + g_1 u_1 \\ y(x,0) &= 0 \\ y(0,t) &= y(a,t) = 0 \end{aligned} \tag{2.4}$$

Since the transport term does not affect the zeros of the eigenfunctions.

2.1.2 Zone controls strategic for two-dimensional domains

Rectangular domain $\Omega =]0, \alpha_1[\times]0, \alpha_z[$

Consider the system

$$\begin{aligned} \frac{\partial y}{\partial t} &= \Delta y + \sum_{i=1}^{p} g_i u_i & & x = x_1, x_2 \in \Omega, \quad t \in]0, T[\\ y(x,0) &= 0 & & \text{in } \Omega \\ y(\xi,t) &= 0 & & \xi = (\xi_1, \xi_2) \in \partial\Omega, \quad t \in]0, T[\end{aligned} \tag{2.5}$$

In this case the eigenfunctions and eigenvalues are

$$\phi_{mn}(x_1, x_2) = \frac{2}{(\alpha_1 \alpha_2)^{1/2}} \sin\left(\frac{m\pi x_1}{\alpha_1}\right) \sin\left(\frac{m\pi x_2}{\alpha_2}\right) \tag{2.6}$$

$$\lambda_{mn} = -\left(\frac{m^2}{\alpha_1^2} + \frac{n^2}{\alpha_2^2}\right)\pi^2 \qquad m, n \geqslant 1 \tag{2.7}$$

The order of multiplicity of the eigenvalues may be infinite and then it is not possible to achieve weak controllability with a finite number of controls. However, if we suppose that

$$\frac{\alpha_1^2}{\alpha_2^2} \notin \mathbb{Q} \tag{2.8}$$

then $r_n = 1$ for all n. Hence, it may be possible to achieve weak controllability with one control. In Ω the support of the control may have many forms. We shall examine two cases which correspond to realistic possibilities.

Proposition 2.3 Controls with rectangular supports

Suppose that

$$\Omega_1 = \prod_{i=1}^{2} [a_i - l_i, a_i + l_i] \subset \Omega = \prod_{i=1}^{2}]0, \alpha_i[$$

then the control (Ω_1, g_1) is not strategic if for some i_0, $1 \leqslant i_0 \leqslant 2$, $a_{i_0}/\alpha_{i_0} \in \mathbb{Q}$ and g_1 is symmetric about the line $x_{i_0} = a_{i_0}$.

Proof Suppose that $i_0 = 1$, $a_1/\alpha_1 \in \mathbb{Q}$, then there exists an integer $m_0 \geqslant 1$, such that $\sin(m_0\pi a_1/\alpha_1) = 0$. Now

$$\langle g_1, \phi_{m_0 m_0} \rangle = \frac{2}{(\alpha_1\alpha_2)^{1/2}} \int_{a_2-l_2}^{a_2+l_2} \left\{ \int_{a_1-l_1}^{a_1+l_1} g_1(x_1, x_2) \sin\left(\frac{m_0\pi x_1}{\alpha_1}\right) \mathrm{d}x_1 \right\} \sin\left(\frac{m_0\pi x_2}{\alpha_2}\right) \mathrm{d}x_2$$

If g_1 is symmetric about the line $x_1 = a_1$, then the integral in curly brackets is zero. Hence $\langle g_1, \phi_{m_0 m_0} \rangle = 0$ and the result follows from Proposition 2.1. ■

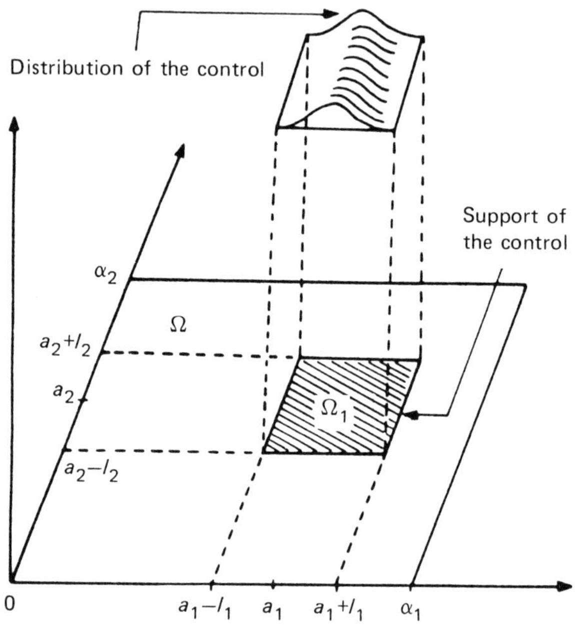

Fig. 2.1 Controls with rectangular supports in rectangular domains

Proposition 2.4 Controls with circular supports

If

$$\Omega_1 = D(c, r) \subset \Omega = \prod_{i=1}^{2}]0, \alpha_i[\qquad c = (c_1, c_2)$$

Then the control (Ω_1, g_1) is not strategic if, for some i_0, $1 \leqslant i_0 \leqslant 2$, $c_{i_0}/\alpha_{i_0} \in \mathbb{Q}$ and g_1 is symmetric about the line $x_{i_0} = c_{i_0}$.

Proof Suppose that $i_0 = 1$, then these exists an integer m_0 such that $\sin(m_0\pi c_1/\alpha_1) = 0$. Now

$$\langle g_1, \phi_{m_0 m_0} \rangle = \frac{2}{(\alpha_1\alpha_2)^{1/2}} \int_0^r \int_0^{2\pi} g_1(c_1 + r\cos\theta, c_2 + r\sin\theta) \sin\left\{\frac{m_0\pi(c_1 + r\cos\theta)}{\alpha_1}\right\}$$
$$\times \sin\left\{\frac{m_0\pi(c_2 + r\sin\theta)}{\alpha_2}\right\} r\,\mathrm{d}r\,\mathrm{d}\theta$$

Since g_1 is symmetric about $x_1 = c_1$, the function

$$(r, \theta) \to g_1(c_1 + r\cos\theta, c_2 + r\sin\theta)\sin\left\{\frac{m_0\pi(c_2 + r\sin\theta)}{\alpha_2}\right\}$$

is symmetric on $[0, \pi]$ about $\theta = \pi/2$ for all r. However, the function

$$(r, \theta) \to \sin\left\{\frac{m_0\pi(c_1 + r\cos\theta)}{\alpha_1}\right\}$$

is antisymmetric on $[0, \pi]$ about $\pi/2$. By decomposing the integral as a sum on $[0, \pi]$ and $[\pi, 2\pi]$, it is easy to see that

$$\langle g_1, \phi_{m_0 m_0}\rangle = 0$$

So the result follows from Proposition 2.1. ■

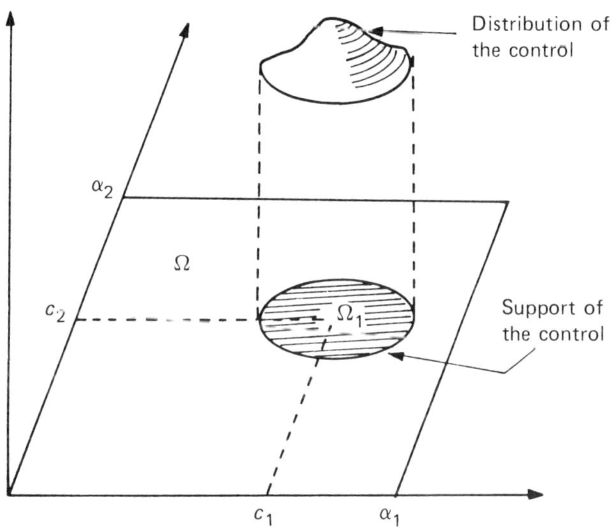

Fig. 2.2 Controls with circular supports in rectangular domains

***Circular domain*: $\Omega = D(0, 1)$**

Consider the system

$$\begin{aligned} &\frac{\partial z}{\partial t}(r, \theta, t) = \Delta z(r, \theta, t) + \sum_{i=1}^{p} g_i(r, \theta)u_i(t) && \text{in } \Omega \times (0, T) \\ &z(r, \theta, 0) = z_0(r, \theta) && \text{in } \Omega \\ &z(1, \theta, t) = 0 && \theta \in [0, 2\pi], \quad t \in (0, T). \end{aligned}$$

The eigenvalues and eigenfunctions are

$$\lambda_{mn} = -\beta_{mn}^2 \qquad n \geqslant 0, \quad m \geqslant 1 \tag{2.9}$$

where β_{mn} are the zeros of the Bessel functions J_n, and

$$\begin{aligned} \phi_{0m}(r,\theta) &= J_0(\beta_{0m}r) && m \geqslant 1 \\ \phi_{nm_1}(r,\theta) &= J_n(\beta_{nm_1} r)\cos(n\theta) && n, m_1 \geqslant 1 \\ \phi_{nm_2}(r,\theta) &= J_n(\beta_{nm_2} r)\sin(n\theta) && n, m_2 \geqslant 1 \end{aligned} \tag{2.10}$$

The order of multiplicity is $r_n = 2$ for all $n, m \neq 0$. $r_n = 1$ for all $n, m = 0$. So the system cannot be weakly controllable with one actuator or weakly observable with one sensor. We consider the case of two actuators $(\Omega_i, g_i), i = 1, 2$, where Ω_i is symmetric about the axis $\theta = \theta_i$. We have the following.

Proposition 2.5

The pair of controls $(\Omega_i, g_i)_{1 \leqslant i \leqslant 2}$ is not strategic if, for each i, $1 \leqslant i \leqslant 2$, g_i is symmetric about $\theta = \theta_i$ and $\theta_i/\pi \in \mathbb{Q}$.

The demonstration of this result is very technical and can be found in the work of Berrahmoune (1984) for rectangular, circular and sectorial supports.

Remark 2.3

The result of Proposition 2.5 is unchanged with result to rotations and so we may replaced the condition $\theta_i/\pi \in \mathbb{Q}$ by $(\theta_1 - \theta_2)/\pi \in \mathbb{Q}$.

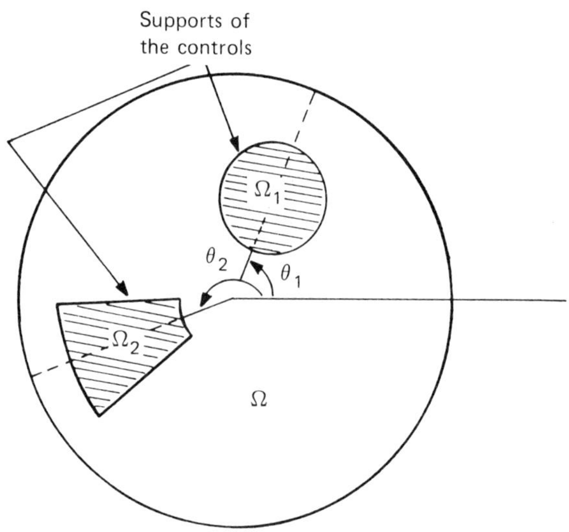

Fig. 2.3 Circular domain

2.2 POINT CONTROLS

Recall the system (S_p) considered in the previous chapter:

$$(S_p) \quad \begin{cases} \dfrac{\partial y}{\partial t} - \Delta y = \sum_{i=1}^{p} \delta(x - b_i)u_i & \text{in } \Omega \times]0, T[\\ y(x, 0) = 0 & \text{in } \Omega \\ y(\xi, t) = 0 & \text{in } \partial\Omega \times]0, T[\end{cases}$$

We understand (S_{p}) as a particular case of (S)

$$(S)\qquad \begin{cases} \dot{z} = Az + Bu \\ z(0) = 0 \end{cases}$$

where, in the case of point controls, (S) is defined on $D(A)'$ $(D(A) \subset L^2(\Omega) \subset D(A)')$ and A generates a strongly continuous semigroup on each of the spaces. For $u \in L^2(0, T; \mathbb{R}^p)$, one need not considers the controllability in $L^2(\Omega)$. Take for example $n = 2$; then

$$\delta(.-b_i) \in H^{-2}(\Omega)$$

and

$$\Delta y \in L^2(0, T; H^{-2}(\Omega)), \qquad \frac{\partial y}{\partial t} \in L^2(0, T; H^{-2}(\Omega))$$

Thus $t \to y(.,t)$ is continuous on $]0, T[$ in $H^{-2}(\Omega)$. Actually, one can show that there is more regularity and $t \to y(.,t)$ is continuous on $]0, T[$ in $H^{-1}(\Omega)$.

One may assume more regularity in the control action to obtain

$$y(.,T) \in L^2(\Omega)$$

This is the case if $u \in \mathscr{D}(0, T)^p$, but such a choice is not very realistic. Instead, we shall choose the controls in $L^r(0, T; \mathbb{R}^p)$ and seek the value of r so that the operator

$$H: u \to Hu = \int_0^T S(T-\tau) Bu(\tau)\, \mathrm{d}\tau$$

maps $L^r(0, T; \mathbb{R}^p)$ into $L^2(\Omega)$.

Recall that, for the case of point controls,

$$B: \mathbb{R}^p \to D(A)'$$

and the weak solution of (S) is given by

$$z_{\mathrm{u}}(t) = \sum_{i=1}^{p} \int_0^t S(t-\tau) \delta_{b_i} u_i(\tau)\, \mathrm{d}\tau \tag{2.11}$$

We shall apply the following result.

Proposition 2.6

Suppose that there exists a Hilbert space W, such that

(1) $L^2(\Omega) \subset W$ with continuous dense injection;
(2) $\delta_{b_i} \in W$, $i = 1, \ldots, p$;
(3) $S(t) \in \mathscr{L}(W, L^2(\Omega))$, for all $t > 0$ and $\| S(t)z \|_{L^2(\Omega)} \leqslant h(t) \| z \|_W$ for all $z \in W$ with $h(.) \in L^s(0, T)$.

Then z_{u} given by equation (2.11) is continuous from $]0, T[$ with values in $L^2(\Omega)$ for all $u \in L^r(0, T; \mathbb{R}^p)$, such that

$$\frac{1}{r} + \frac{1}{s} = 1$$

For the proof, see Curtain and Pritchard (1978). ■

For the application to our case, we let

$$W = H^{-1-\varepsilon}(\Omega) \qquad 0 < \varepsilon < 1 - \frac{2}{r}$$

Since $S(t) \in \mathscr{L}(L^2(\Omega), H^k(\Omega))$ for all $k \geqslant 0$, we have as an application of certain results on interpolation spaces (Lions and Magenes, (1968)

$$\| S(t)z \|_{H^{1+\varepsilon}(\Omega)} \leqslant M \| S(t)z \|_{H^2(\Omega)}^{(1+\varepsilon)/2} \| S(t)z \|_{L^2(\Omega)}^{(1-\varepsilon)/2}$$
$$\leqslant M \| S(t)z \|_{H^2(\Omega)}^{(1+\varepsilon)/2} \| z \|_{L^2(\Omega)}^{(1-\varepsilon)/2}$$

for some generic constant M.

Now

$$S(t)z = \sum_n e^{\lambda_n t} \sum_{j=1}^{r_n} \langle z, \phi_{nj} \rangle \phi_{nj}$$

Therefore

$$\| S(t)z \|_{H^2(\Omega)}^2 = \sum_n e^{2\lambda_n t} \sum_{j=1}^{r_n} \langle z, \phi_{nj} \rangle^2 \| \phi_{nj} \|_{H^2(\Omega)}^2$$

but

$$\| \phi_{nj} \|_{H^2(\Omega)}^2 \leqslant M |\lambda_n|^2$$

(Lions and Magenes, 1968) and hence

$$\| S(t)z \|_{H^2(\Omega)}^2 \leqslant \frac{M}{t^2} \| z \|_{L^2(\Omega)}^2$$

So

$$\| S(t)z \|_{H^{1+\varepsilon}(\Omega)} \leqslant M \left(\frac{1}{t} \right)^{(1+\varepsilon)/2} \| z \|_{L^2(\Omega)}$$

Thus we may choose $h(t) = (1/t)^{(1+\varepsilon)/2}$ with $s < 2/(1+\varepsilon)$. This leads to the choice $r > 2/(1-\varepsilon)$.

Proposition 2.7

The suite of point controls $(b_i, \delta_{b_i})_{1 \leqslant i \leqslant p}$ is strategic for (S_p) if and only if

(1) $p \geqslant \sup_n (r_n)$;

(2) $\operatorname{rank}(G_n) = r_n$ for all n with $(G_n)_{ij} = \phi_{nj}(b_i)$, $1 \leqslant i \leqslant p$, $1 \leqslant j \leqslant r_n$.

The proof is analogous to that of Proposition 1.2. Let us look at some applications.

2.2.1 Point controls strategic for one-dimensional domains

Consider the case where $\Omega =]0, a[$, (ϕ_n) are given by equation (2.1) and λ_n by equation (2.2). Since for all $n, r_n = 1$, one control may be sufficient to ensure weak controllability of (S_p). More precisely, we consider the system

$$\begin{aligned}
&\frac{\partial y}{\partial t} - \Delta y = \delta(x - b_1)u && \text{in }]0, a[\times]0, T[\\
&y(0, t) = y(a, t) = 0 && t \in]0, t[\qquad (2.12) \\
&y(x, 0) = 0 && x \in]0, a[
\end{aligned}$$

For some examples of systems excited by these types of control, see Babary (1978) and Hariri *et al.* (1985).

Proposition 2.8

The control (b_1, δ_{b_1}) is strategic for the system (2.12) if and only if $b_1/a \notin \mathbb{Q}$.

2.2.2 Point controls strategic for two-dimensional domains

Rectangular domain: $\Omega =]0, \alpha_1[\times]0, \alpha_2[$

The eigenfunctions and eigenvalues for this case are given by equations (2.6) and (2.7). Under the condition (2.8), one control may be sufficient to ensure weak controllability of (S_p). We have the following.

Proposition 2.9

If $b = (b_1, b_2) \in \Omega$, then the control (b, δ_b) is strategic if and only if $b_i/\alpha_i \notin \mathbb{Q}$, $i = 1, 2$.

Circular domain: $\Omega = D(0, 1)$

Recall that in this case we require at least two controls. Let us take $p = 2$ and consider two point controls located at polar coordinates $c_1 = (\rho_1, \theta_1)$, $c_2 = (\rho_2, \theta_2)$.

Proposition 2.10

(1) The pair of controls $(c_i, \delta_{c_i})_{1 \leqslant i \leqslant 2}$ is not strategic if $(\theta_1 - \theta_2)/\pi \in \mathbb{Q}$.
(2) If $\rho_1 = \rho_2$, then the point controls $(c_1, \delta_{c_i})_{1 \leqslant i \leqslant 2}$ are strategic if and only if $(\theta_1 - \theta_2)/\pi \notin \mathbb{Q}$.

Remark 2.4

Controls of the filament type may be important in this framework. Fig. 2.4 represents a control of this type where the support is one dimensional. More generally, suppose that γ is an injective immersion of an open set θ in $\mathbb{R}^l$ into Ω with

$$\gamma^{-1}: \Omega_1 = \gamma(\theta) \to \theta$$

One says that Ω_1 is a regular open set (it is even a subvariety of dimension l). Consider now the system described by

$$\begin{aligned} &\frac{\partial y}{\partial t} - \Delta y = \delta x - \gamma(.)u && \text{in } \Omega \times]0,T[\\ &y(x,0) = 0 && \text{in } \Omega \\ &y(\xi,0) = 0 && \text{in } \partial\Omega \times]0,T[\end{aligned} \tag{2.13}$$

It is easy to see that one may apply the characterisation of weak controllability of Proposition 2.7 with

$$(G_n)_{ij} = \int_{\theta} \phi_{nj}\gamma(s)\,\mathrm{d}s \qquad (p=1)$$

In the case where $n=2$ and $l=1$ (Fig. 2.4) we then have the following.

Corollary

If $\mathrm{Im}(\gamma) = \Omega_1$ is symmetric about the line $x = x_0$, then the filament control with support Ω_1 is not strategic if $x_0/\alpha_1 \in \mathbb{Q}$.

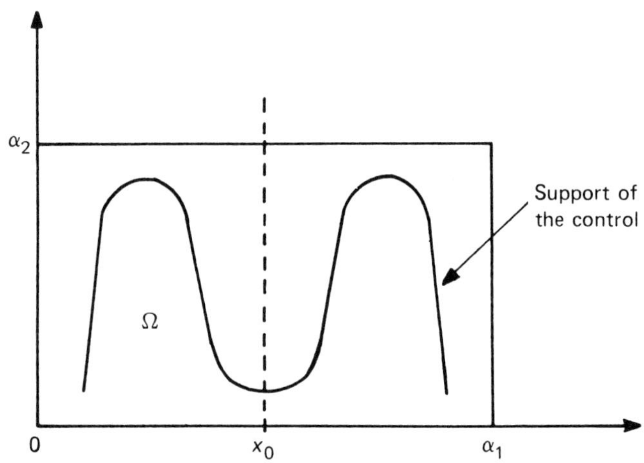

Fig. 2.4 Filament control

This type of control is important for diffusion systems excited by heating filaments.

2.3 BOUNDARY CONTROLS

We shall consider in this section linear parabolic systems excited by zone or point controls which are located in the boundary of the system. This corresponds to many physical problems. For example a turning furnace may only be heated at the boundary (Fig. 2.5).

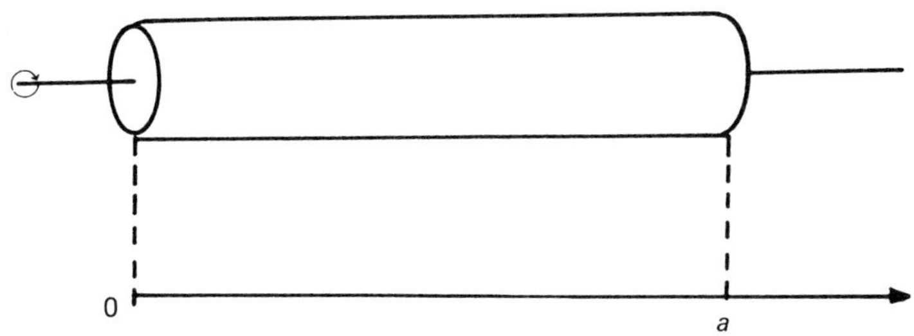

Fig. 2.5

In order of increasing difficulty the case of boundary controls is the highest. Consider the two systems

$$(S_f) \quad \begin{cases} \dfrac{\partial y}{\partial t} - \Delta y = 0 & \text{in } \Omega \times]0, T[\\ y(x, 0) = 0 & \text{in } \Omega \\ y(\xi, t) = \sum_{i=1}^{p} g_i(\xi) u_i(t) & \text{in } \partial\Omega \times]0, T[\end{cases}$$

(S_f) corresponds to a system excited by p zone controls localized on the boundary at $\Gamma_i \subset \partial\Omega$ with $g_i \in L^2(\Gamma_i)$.

$$(S'_f) \quad \begin{cases} \dfrac{\partial y}{\partial t} - \Delta y = 0 & \text{in } \Omega \times]0, T[\\ y(x, 0) = 0 & \text{in } \Omega \\ y(\xi, t) = \sum_{i=1}^{p} \delta(\xi - b_i) u_i(t) & \text{in } \partial\Omega \times]0, T[\end{cases}$$

(S'_f) corresponds to the case of p point controls localized in the boundary $\partial\Omega$ at points b_i.

In both cases, we may also consider Neumann boundary conditions.

The difficulty that we find in both cases in that of determining hypotheses such that (S_f) and (S'_f) have unique solutions. In general, $y(., T)$ will not be in $L^2(\Omega)$ so that, as in section 2.2, it may be possible to assume more regularity in the control u.

It is convenient to distinguish two cases.

2.3.1 Zone controls in the boundary

If $Az = \Delta z$, $z \in D(A) = H_0^1(\Omega) \cap H^2(\Omega)$, then A generates a strongly continuous analytic semigroup $(S(t))_{t \geq 0}$ and (S_f) admits a weak solution

$$y_u(t) = -\sum_{i=1}^{p} A \int_0^t S(t - \tau) G g_i u_i(\tau) \, d\tau \tag{2.14}$$

In equation (2.14), G is a Green's operator defined by

$$G : L^2(\partial\Omega) \to L^2(\Omega)$$
$$h \to p$$

where

$$\Delta p = 0 \qquad \text{in } \Omega$$
$$p|_{\partial\Omega} = h$$

The operator H given by

$$H: L^2(0, T; \mathbb{R}^p) \rightarrow L^2(0, T; L^2(\Omega))$$
$$u \rightarrow y_u$$

is continuous, and $y_u(t) \in L^2(\Omega)$ almost everywhere on $]0, T[$ but we cannot be sure that $y_u(T) \in L^2(\Omega)$. We apply the following result to ensure that $y_u(T) \in L^2(\Omega)$.

Proposition 2.11

Under the preceding hypothesis if $u \in L^r(0, T; \mathbb{R}^p)$ with $r > 4$, then $y_u(T) \in L^2(\Omega)$.

The proof of this result makes use of the abstract theory of non-homogeneous boundary value problems and relies on the result that

$$\| AS(t)G \| \leqslant \gamma(t) \qquad \text{almost everywhere in }]0, T[$$

with $\gamma(t) = Mt^{-3/4}$ (Washburn, 1976).

Remark 2.5

In the case where Neumann boundary conditions are considered, the same problems arise but then one finds that, for all $u \in L^2(0, T; \mathbb{R}^p)$, $y_u(T) \in L^2(\Omega)$.

Therefore, in the rest of this section, we take

$$u \in L^r(0, T; \mathbb{R}^p) \qquad r > 4 \tag{2.15a}$$

for the Dirichlet problem and

$$u \in L^2(0, T; \mathbb{R}^p) \tag{2.15b}$$

for the Neumann problem.

Suppose that (ϕ_{nj}) are the eigenfunctions for the Dirichlet problem (and (ψ_{nj}) the eigenfunctions for the Neumann problem) and λ_n are the eigenvalues with multiplicity r_n. We have the following.

Proposition 2.12

The suite of zone controls on the boundary $(\Gamma_i, g_i)_{1 \leqslant i \leqslant p}$ are strategic for (S_f) if and only if

(1) $p \geqslant \sup_n (r_n)$;

(2) $\operatorname{rank}(G_n) = r_n$ for all n where $(G_n)_{ij} = \langle g_i, (\partial\phi_{nj}/\partial\nu) \rangle_{L^2(\Gamma_i)}$ for Dirichlet problems and $(G_n)_{ij} = \langle g_i, \psi_{nj} \rangle_{L^2(\Gamma_i)}$ for Neumann problems, $i = 1, \ldots, p$, $j = 1, \ldots, r_n$.

Proof Consider the Dirichlet case. Then the suite of controls which is strategic is equivalent to

$$\langle y_u(T), h \rangle = 0 \qquad \forall u \in L^r(0, T; \mathbb{R}^p) \Rightarrow h = 0$$

Consider the equations

$$-\frac{\partial\Psi}{\partial t}-\Delta\Psi=0 \qquad \text{in } \Omega\times]0,T[$$

$$\Psi(x,T)=h \qquad \text{in } \Omega$$

$$\Psi(\xi,t)=0 \qquad \text{in } \partial\Omega\times]0,T[$$

Then we have

$$\langle y_u(T),h\rangle = -\int_{\partial\Omega\times]0,T[}\left(\sum_{i=1}^{p} g_i u_i\right)\frac{\partial\Psi}{\partial\nu}\,d\xi\,dt \tag{2.16}$$

Also

$$\Psi(t)=S(T-t)h$$

where

$$S(T-t)h=\sum_n e^{\lambda_n(T-t)}\sum_{j=1}^{r_n}\langle h,\phi_{nj}\rangle\phi_{nj}$$

Hence,

$$\langle y_u(T),h\rangle=0, \qquad \forall u\in L^r(0,T;\mathbb{R}^p)\Leftrightarrow$$

$$\sum_{i=1}^{p}\int_0^T u_i(t)\sum_n e^{\lambda_n(T-t)}\sum_{j=1}^{r_n}\langle h,\phi_{nj}\rangle\int_{\Gamma_i} g_i\frac{\partial\phi_{nj}}{\partial\nu}\,d\xi\,dt=0$$

for all $u\in L^r(0,T;\mathbb{R}^p)$. This implies that

$$\sum_n e^{\lambda_n(T-t)}\sum_{j=1}^{r_n}\langle h,\phi_{nj}\rangle\int_{\Gamma_i} g_i\frac{\partial\phi_{nj}}{\partial\nu}\,d\xi=0$$

for all $t\in]0,T[$, which in turn implies that

$$\sum_{j=1}^{r_n}\langle h,\phi_{nj}\rangle\int_{\Gamma_i} g_i\frac{\partial\phi_{nj}}{\partial\nu}\,d\xi=0$$

for all $i=1,\ldots,p$, $n\in\mathbb{N}$. Thus, $\langle y_u(T),h\rangle=0\Rightarrow h=0$ is equivalent to

$$G_nV=0\Rightarrow V=0 \qquad \forall n$$

where

$$V=[\langle h,\phi_{n1}\rangle\cdots\langle h,\phi_{n\Gamma_n}\rangle]^{\mathrm{T}}$$

from which the result follows. For the Neumann case the analogue of equation (2.16) is

$$\langle y_u(T),h\rangle=\int_{\partial\Omega\times]0,T[}\left(\sum_{i=1}^{r} g_i u_i\right)\psi\,d\xi\,dt \tag{2.17}$$

2.3.2 Zone boundary controls for two-dimensional domains

Consider the case where $\Omega=]0,\alpha_1[\times]0,\alpha_2[$ with condition (2.8) so that one control may be sufficient for the controllability of (S_f). The system is

$$\begin{aligned}
&\frac{\partial y}{\partial t}-\Delta y=0 &&\text{in } \Omega\times]0,T[\\
&y(x,0)=0 &&\text{in } \Omega\\
&y(\xi,t)=g_1(\xi)u_1(t) &&\text{in } \partial\Omega\times]0,T[
\end{aligned} \tag{2.18}$$

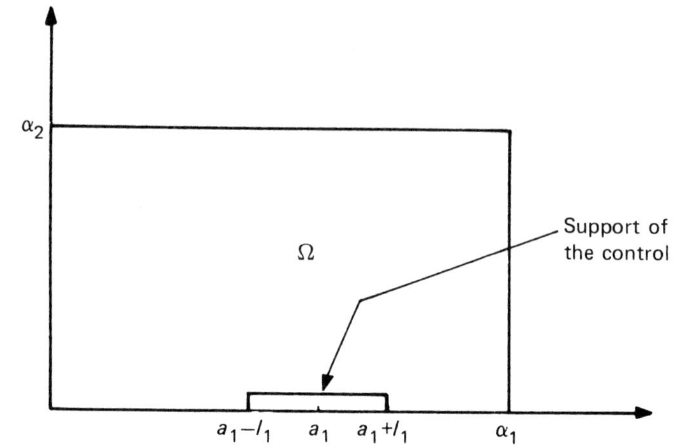

Fig. 2.6 $\Gamma_1=[a_1-l_1,a_1+l_1]\times\{0\}$

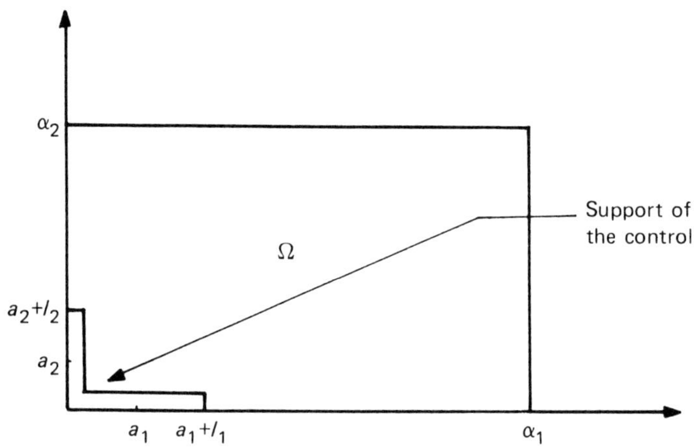

Fig. 2.7 $\Gamma_2=[0,a_1+l_1]\times\{0\}\cup\{0\}\times[0,a_2+l_2]$

For the support of the control Γ_1, we consider two cases. Case 1 is on one side of the boundary (Fig. 2.6):

$$\Gamma_1=[a_1-l_1,a_1+l_1]\times\{0\}$$

Case 2 is on two sides of the boundary (Fig. 2.7):

$$\Gamma_2=[0,a_1+l_1]\times\{0\}\cup\{0\}\times[0,a_2+l_2]$$

By applying Proposition 2.12, we have the following results.

Proposition 2.13 Case 1 (Fig. 2.6)

Suppose that g_1 is symmetric about the line $x_1=a_1$; then

(1) the control (Γ_1,g_1) is not strategic for the system (2.18) if $a_1/\alpha_1\in\mathbb{Q}$;
(2) the control (Γ_1,g_1) is not strategic for the system (2.18) with Neumann boundary conditions if there exists an integer m such that $2a_1m/\alpha_1$ in odd.

Proposition 2.14 Case 2 (Fig. 2.7)

Suppose that g_2 is doubly symmetric about $x_1 = a_1$ and $x_2 = a_2$; then

(1) the control (Γ_2, g_2) is not strategic for the system (2.18) if $a_i/\alpha_i \in \mathbb{Q}, i = 1, 2$;
(2) the control (Γ_2, g_2) is not strategic for the system (2.18) with Neumann boundary conditions if there exist two integers m and n such that $2a_1 m/\alpha_1$ and $2a_2 n/\alpha_2$ are odd.

Proof Suppose that we have Dirichlet boundary conditions.

For case 1 if $a_1/\alpha_1 \in \mathbb{Q}$, there exists an integer m such that

$$\int_{a_1 - l_1}^{a_1 + l_1} g_1(x_1, 0) \sin\left(\frac{m\pi x_1}{\alpha_1}\right) dx_1 = 0$$

For case 2, if $a_i/\alpha_i \in \mathbb{Q}$, $i = 1, 2$, there exist integers m and n such that

$$\frac{m\pi}{\alpha_1} \int_0^{a_2 + l_2} g_2(0, x_2) \sin\left(\frac{n\pi x_2}{\alpha_2}\right) dx_2 - \frac{n\pi}{\alpha_1} \int_0^{a_1 + l_1} g_2(x_1, 0) \sin\left(\frac{m\pi x_1}{\alpha_1}\right) dx_1 = 0$$

The Neumann case can be proved in a similar way. ■

Remark 2.6

Cases 1 and 2 have been chosen as typical examples. The results of Propositions 2.13 and 2.14 are also valid for controls with other types of support.

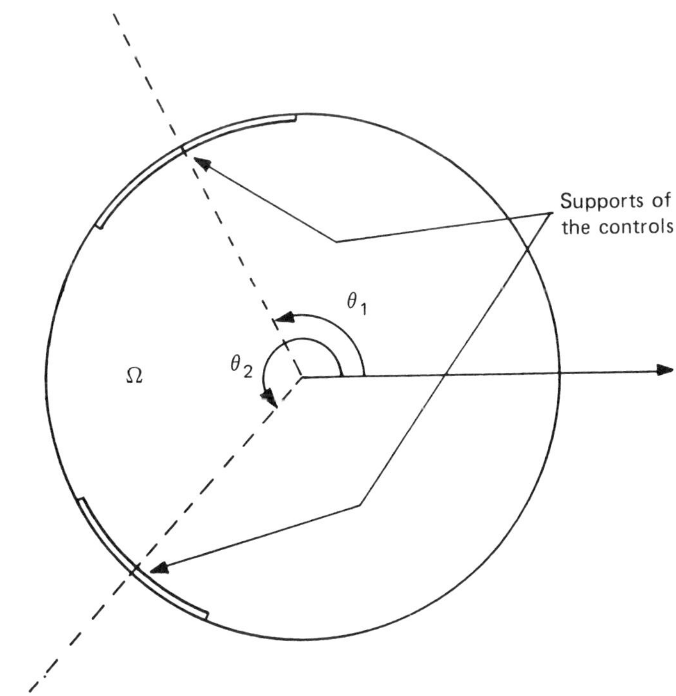

Fig. 2.8 Zone controls on the boundary

Remark 2.7

In the case of a circular domain $\Omega = D(0,1)$, then it is necessary to have two controls. One can easily show that, if g_i are symmetric about θ_i, $i = 1, 2$, then the pair of controls $(\Gamma_i, g_i)_{1 \leqslant i \leqslant 2}$ is not strategic if $(\theta_1 - \theta_2)/\pi \in \mathbb{Q}$ (Fig. 2.8).

2.3.3 Point controls on the boundary

Consider the system (S'_f)

$$(S'_f) \qquad \begin{cases} \dfrac{\partial y}{\partial t} - \Delta y = 0 & \text{in} \quad \Omega \times]0, T[\\ y(x, 0) = 0 & \text{in} \quad \Omega \\ y(\xi, t) = \sum\limits_{i=1}^{p} \delta(\xi - b_i) u_i(t) & \text{in} \quad \partial\Omega \times]0, T[\end{cases}$$

This system is excited by p point controls located in the boundary at points b_i.

The solution of (S'_f) may be defined by the method of transposition. Suppose that ψ is given and ϕ solves the equations

$$\begin{aligned} -\frac{\partial \phi}{\partial t} - \Delta \phi = \psi \qquad & \text{in} \quad \Omega \times]0, T[\\ \phi(x, T) = 0 \qquad & \text{in} \quad \Omega \\ \phi(\xi, t) = 0 \qquad & \text{in} \quad \partial\Omega \times]0, T[\end{aligned}$$

One multiplies (S'_f) by ϕ and, after a formal application of Green's formula,

$$\int_{\Omega \times]0, T[} y\psi \,\mathrm{d}x\,\mathrm{d}t = -\sum_{i=1}^{p} \int_0^T \frac{\partial \phi(b_i, t)}{\partial \nu} u_i(t)\,\mathrm{d}t \tag{2.20}$$

One utilizes equation (2.20) in the definition of a solution of (S'_f) (Lions and Magenes, 1968; Lions, 1981). For $u = [u_1, \ldots, u_p] \in L^2(0, T; \mathbb{R}^p)$ $y(., T)$ need not lie in $L^2(\Omega)$. Suppose that we are considering the case where $n = 3$. If $\psi \in L^2(0, T; H_0^\beta)$ with $\beta > \frac{1}{2}$, then the solution ϕ of equation (2.19) lies in $L^2(0, T; H^{\beta+2}(\Omega) \cap H_0^1(\Omega))$ and hence by the trace theorem

$$\frac{\partial \phi}{\partial \nu} \in L^2(0, T; H^{\beta + 1/2}(\partial\Omega))$$

Since $n = 3$ and $\beta > \frac{1}{2}$, we have

$$H^{\beta + 1/2}(\partial\Omega) \subset C^0(\partial\Omega)$$

and so

$$\frac{\partial \phi(b_i, t)}{\partial \nu} \in L^2(0, T)$$

for all i, $1 \leqslant i \leqslant p$. As a result of equation (2.20) we find that $y \in L^2(0, T; H^{-\beta}(\Omega))$ and (S'_f) gives $(\partial y/\partial t) \in L^2(0, T; H^{-\beta-2}(\Omega))$. Then $y(., T)$ is defined in $H^{-\beta-1}(\Omega)$ (Lions, 1980b, 1981). In order to obtain $y(., T) \in L^2(\Omega)$, we need to impose more regularity on the controls.

Proposition 2.15

Suppose that Ω is open and bounded in $\mathbb{R}^n$ with $n \leqslant 3$; if $u \in (C_0^\infty(0, T))^p$, then $y(., T) \in L^2(\Omega)$.

Proof Suppose that (ϕ_{nj}) are the eigenfunctions of Δ with Dirichlet boundary conditions, and λ_n are the corresponding eigenvalues.

Set

$$\psi = h(t)\phi_{nj} \qquad \text{with} \quad h(T) = 0$$

Then

$$\begin{aligned}\int_{\Omega\times]0,T[} \left(\frac{\partial y}{\partial t} - \Delta y\right)\psi \,\mathrm{d}x\,\mathrm{d}t &= \int_{\Omega\times]0,T[} y\left(-\frac{\partial\psi}{\partial t} - \Delta\psi\right)\mathrm{d}x\,\mathrm{d}t + \int_{\partial\Omega\times]0,T[} y\frac{\partial\psi}{\partial v}\,\mathrm{d}\xi\,\mathrm{d}t \\ &= \int_{\Omega\times]0,T[} y(-h' - \lambda_n h)\phi_{nj}\,\mathrm{d}x\,\mathrm{d}t + \sum_{i=1}^{p} \frac{\partial\phi_{nj}(b_i)}{\partial v}\int_0^T hu_i\,\mathrm{d}t \\ &= 0\end{aligned}$$

For

$$y = \sum_n \sum_{j=1}^{r_n} z_{nj}(t)\phi_{nj}$$

one has

$$\dot{z}_{nj} - \lambda_n z_{nj} = \sum_{i=1}^{p} \frac{\partial\phi_{nj}(b_i)u_i}{\partial v}$$

$$z_{nj}(0) = 0$$

Hence

$$\begin{aligned}z_{nj}(T) &= \sum_{i=1}^{p} \frac{\partial\phi_{nj}(b_i)}{\partial v}\int_0^T \mathrm{e}^{\lambda_n(T-t)}u_i(t)\,\mathrm{d}t \\ &= -\sum_{i=1}^{p} \frac{\partial\phi_{nj}(b_i)}{\partial v}\mathrm{e}^{\lambda_n T}\int_0^T \frac{\mathrm{d}^k}{\mathrm{d}t^k}\frac{\mathrm{e}^{-\lambda_n t}}{(-\lambda_n)^k}u_i(t)\,\mathrm{d}t \\ &= (-1)^{k+1}\sum_{i=1}^{p} \frac{\partial\phi_{nj}(b_i)}{\partial v}\frac{\mathrm{e}^{\lambda_n T}}{\lambda_n^k}\int_0^T \mathrm{e}^{-\lambda_n t}u_i^{(k)}(t)\,\mathrm{d}t\end{aligned}$$

Since $\phi \to (\partial\phi/\partial v)$ is continuous on $H^s, s > \frac{5}{2}$, one has

$$\left|\frac{\partial\phi_{nj}(b_i)}{\partial v}\right| \leqslant M\|\phi_{nj}\|_{H^s(\Omega)} \leqslant M|\lambda_n|^{s/2}$$

and so

$$|z_{nj}(T)| \leqslant M|\lambda_n|^{(s/2)-k}$$

Thus, for k sufficiently large, we have

$$\|y(., T)\|^2_{L^2(\Omega)} \leqslant \sum_n \sum_{j=1}^{r_n} |z_{nj}(T)|^2 < \infty$$

■

Remark 2.8

In the case of Neumann boundary conditions, one can show by the same approach that $y(.,T)\in H^{-1}(\Omega)$ for $u\in L^2(0,T,\mathbb{R}^p)$. For $u\in(C_0^\infty(0,T)^p)$, one has the same result as that of Proposition 2.15.

We have the following characterization.

Proposition 2.16

The suite of point controls on the boundary $(b_i,\delta_{b_i})_{1\leqslant i\leqslant p}$ is strategic for the system (S_f') if and only if

(1) $p\geqslant\sup_n(r_n)$;

(2) rank $(G_n)=r_n$ for all n where $(G_n)_{ij}=(\partial\phi_{nj}(b_i)/\partial v)$ for the Dirichlet case and $(G_n)_{ij}=\psi_{nj}(b_i)$ for the Neumann case, $i=1,\dots,p$, $j=1,\dots,r_n$.

Proof For $f\in L^2(\Omega)$, consider

$$-\frac{\partial\phi}{\partial t}-\Delta\phi=0 \qquad \text{in} \quad \Omega\times]0,T[$$

$$\phi(x,T)=f \qquad \text{in} \quad \Omega$$

$$\phi(\xi,t)=0 \qquad \text{in} \quad \partial\Omega\times]0,T[$$

With $u=[u_1,u_2,\dots,u_p]\in(C_0^\infty(0,T)^p$, we have via Green's formula

$$\int_{\Omega\times]0,T[}\left(\frac{\partial y}{\partial t}-\Delta y\right)\phi\,\mathrm{d}x\,\mathrm{d}t=\langle y(.,T),f\rangle+\sum_{i=1}^{p}\int_{\partial\Omega\times]0,T[}u_i\delta_{b_i}(\xi)\frac{\partial\phi}{\partial v}\,\mathrm{d}\xi\,\mathrm{d}t$$

However,

$$\int_{\partial\Omega\times]0,T[}u_i\delta_{b_i}(\xi)\frac{\partial\phi}{\partial v}\,\mathrm{d}\xi\,\mathrm{d}t=\int_0^T u_i(t)\frac{\partial\phi(b_i,t)}{\partial v}\,\mathrm{d}t$$

Hence

$$\langle y(.,T),f\rangle=0\Leftrightarrow\frac{\partial\phi(b_i,t)}{\partial v}=0 \qquad \forall t,\quad \forall i$$

By developing the solution in terms of the eigenfunctions ϕ_{nj} and using a similar argument to the one used for zones, one obtains the result. ■

2.3.4 Applications to particular domains

Case where $\Omega=]0,a[$

By the same considerations as those in section 2.1, one control is sufficient and we have the following.

Proposition 2.17

As b_1 is equal to 0 or a, the point control on the boundary (b_1,δ_{b_i}) is strategic for (S_f') in both the Dirichlet and the Neumann cases.

Proof In the Dirichlet case,

$$\frac{\partial \phi_n(b_1)}{\partial x} = \left(\frac{2}{a}\right)^{1/2} \frac{\partial}{\partial x}\left\{\sin\left(\frac{n\pi x}{a}\right)\right\}\bigg|_{x=b_1} \neq 0 \qquad \text{for } b_1 = 0 \text{ or } a$$

For the Neumann case,

$$\psi_n(b_1) = \left(\frac{2}{a}\right)^{1/2} \cos\left(\frac{n\pi b_1}{a}\right) \neq 0 \qquad \text{for } b_1 = 0 \text{ or } a$$

■

Case where $\boldsymbol{\Omega =]0, \alpha_1[\times]0, \alpha_2[}$

With the condition (2.8), i.e. $\alpha_1^2/\alpha_2^2 \notin \mathbb{Q}$, we can envisage controllability with one control.

Proposition 2.18

Let $b_1 = (a_1, a_2)$; then the point control on the boundary (b_1, δ_{b_1}) is strategic for the system (S_f') if and only if

(1) $a_i/\alpha_i \notin \mathbb{Q}$ for all $i = 1, 2$ in Dirichlet case;
(2) $2ma_i/\alpha_i$ is not odd for $i = 1, 2$ in the Neumann case.

The proof is immediate from Proposition 2.14.

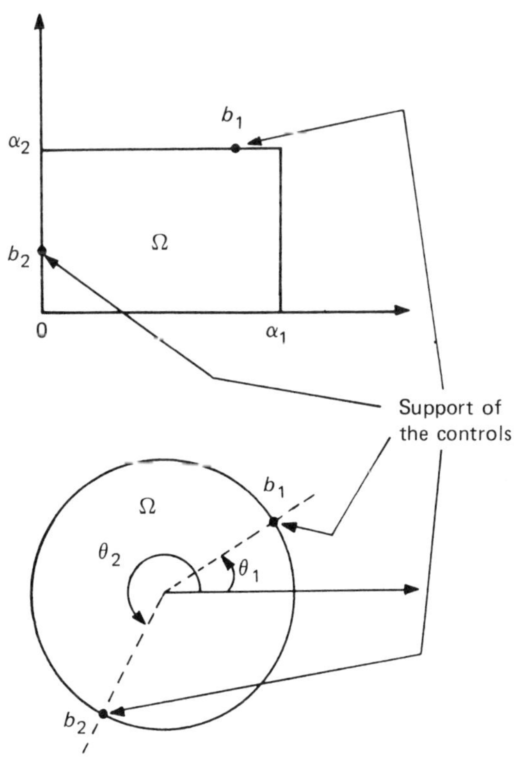

Fig. 2.9 Point controls on the boundary

***Case where* $\Omega = D(0,1)$**

In this case, we need at least two controls. Suppose that these have polar coordinates $b_i = (1, \theta_i)$, $i = 1, 2$, then we have the following.

Proposition 2.19

The pair of controls (b_i, δ_{b_i}) are strategic if and only if $(\theta_1 - \theta_2)/\pi \notin \mathbb{Q}$.

2.4 CASE OF UNBOUNDED DOMAINS

Again we consider the system

$$(S) \qquad \begin{cases} \dot{z} = Az + Bu \\ z(0) = 0 \end{cases}$$

where $Az = \Delta z$, $D(A) = \{z \in L^2(\mathbb{R}^n) : \Delta z \in L^2(\mathbb{R}^n)\}$. Suppose that (S) is excited by p zone controls $(\Omega_i, g_i)_{1 \leqslant i \leqslant p}$ with Ω_i bounded and $g_i \in L^2(\Omega_i)$. We have the following result.

Proposition 2.20

Suppose that $n = 1$; then we have the following.

(1) For the suite $(\Omega_i, g_i)_{1 \leqslant i \leqslant p}$ to be strategic, it is necessary that $p \geqslant 2$.
(2) If $p = 2$, the pair $(\Omega_i, g_i)_{1 \leqslant i \leqslant 2}$ is strategic if and only if

$$\mathscr{F} g_1(x) \mathscr{F} g_2(-x) - \mathscr{F} g_1(-x) \mathscr{F} g_2(x) \neq 0 \qquad \text{almost everywhere in } \mathbb{R}_+ \tag{2.21}$$

where

$$\mathscr{F} g(x) = \int_{\mathbb{R}} \mathrm{e}^{-\mathrm{i}\xi x} g(\xi)\, \mathrm{d}\xi$$

Proof One has

$$\mathscr{F} Az = -x^2 \mathscr{F} z(x) \qquad \forall z \in D(A)$$

and

$$\mathscr{F} S(t) g_i(x) = \mathrm{e}^{-x^2 t} \mathscr{F} g_i(x)$$

Since in this case the semigroup $(S(t))_{t \geqslant 0}$ is analytic,

$$A^n S(t) g_i = S\left(\frac{t}{2}\right) A^n S\left(\frac{t}{2}\right) g_i \qquad \in D(A) \quad \forall t > 0$$

Thus

$$\mathscr{F} A^n S(t) g_i(x) = (-1)^n x^{2n} \mathrm{e}^{-x^2 t} \mathscr{F} g_i(x)$$

The system (S) is weakly controllable if and only if

$$\langle A^n S(t) g_i, z \rangle = 0 \qquad \forall n, \quad i = 1, \ldots, p \Rightarrow z = 0$$

However, this is equivalent to

$$\langle \mathscr{F} A^n S(t) g_i, \mathscr{F} z \rangle = 0 \qquad \forall n, \quad i = 1, \ldots, p \Rightarrow z = 0$$

Now

$$\langle \mathscr{F} A^n S(t) g_i, \mathscr{F} z \rangle = (-1)^n \int_{\mathbb{R}} x^{2n} e^{-x^2 t} \mathscr{F} g_i(x) \overline{\mathscr{F} z(x)} \, dx$$

$$= (-1)^n \int_{\mathbb{R}_+} x^{2n} e^{-x^2 t} [\mathscr{F} g_i(x) \overline{\mathscr{F} z(x)} + \mathscr{F} g_i(-x) \overline{\mathscr{F} z(-x)}] \, dx$$

Since the family (f_n) defined by

$$f_n(x) = x^{2n} e^{-x^2 t} \qquad n > 0, \quad x \in]0, \infty[$$

is dense in $L^2(0, \infty)$, (S) is weakly controllable if and only if

$$\mathscr{F} g_i(x) \overline{\mathscr{F} z(x)} + \mathscr{F} g_i(-x) \overline{\mathscr{F} z(-x)} = 0 \qquad \text{almost everywhere in }]0, \infty[$$

for all $i = 1, \ldots, p$ implies that

$$\mathscr{F} z = 0$$

Thus, if $p = 1$, (S) is not controllable whereas, if $p = 2$, (S) is controllable if

$$\begin{bmatrix} \mathscr{F} g_1(x) & \mathscr{F} g_1(-x) \\ \mathscr{F} g_2(x) & \mathscr{F} g_2(-x) \end{bmatrix} \begin{bmatrix} \mathscr{F} z(x) \\ \mathscr{F} z(-x) \end{bmatrix} = 0 \Rightarrow \mathscr{F} z = 0$$

This gives the condition (2.21). ■

Remark 2.9

If $\Omega = \mathbb{R}^n$ with $n \geqslant 2$, then there does not exist a finite set of strategic controls. In effect, for $n \geqslant 2$,

$$\mathscr{F} S(t) g_i(x) = e^{-\|x\|^2 t} \mathscr{F} g_i(x)$$

and

$$\mathscr{F} A^n S(t) g_i(x) = (-1)^n \| x \|^{2n} \mathscr{F} S(t) g_i(x)$$

Then the suite (f_n) defined by

$$f_n(x) = \| x \|^{2n} e^{-\|x\|^2 t} \forall n$$

is not dense in $L^2(\mathbb{R}^n_+)$ (Szego, 1959).

Corollary

Suppose that $n = 1$ and $p = 2$ and consider the pair of controls $(\Omega_i, g_i)_{1 \leqslant i \leqslant 2}$ where

$$\Omega_i = [x_i - l_i, x_i + l_i], \qquad g_i = \alpha_i \chi_{\Omega_i} \qquad i = 1, 2$$

then the pair is strategic if $x_1 \neq x_2$.

Proof It suffices to see that, in this case, equation (2.21) becomes

$$\mathscr{F} g_1(x) \mathscr{F} g_2(-x) - \mathscr{F} g_1(-x) \mathscr{F} g_2(x) = -\frac{8i\alpha_1\alpha_2}{x^2} \sin(xl_1) \sin(xl_2) \sin\{x(x_1 - x_2)\}$$

from which the result follows. ■

Remark 2.10

In the case where $\Omega =]0, \infty[$ then (*S*) may be controllable with only one control. More precisely, we have the following.

Proposition 2.21

If $\Omega =]0, \infty[$, the zone control (Ω_1, g_1) is strategic for (*S*) if and only if

$$\tilde{\mathscr{F}} g_1(x) = \frac{2}{\pi} \int_{\mathbb{R}_+} \sin(\xi x)\, g_1(\xi)\, \mathrm{d}\xi \neq 0 \qquad \text{almost everywhere in }]0, \infty[\tag{2.22}$$

Proof In the same way that we proved Proposition 2.20 utilizing the Fourier sine transform, we have that (*S*) is controllable if and only if for all $z \in L^2(0, \infty)$

$$\tilde{\mathscr{F}} g_1(x) \tilde{\mathscr{F}} z(x) = 0 \qquad \text{almost everywhere in }]0, \infty[\Rightarrow \tilde{\mathscr{F}} z(x) = 0$$

This requires that $\tilde{\mathscr{F}} g_1(x) \neq 0$ almost everywhere in $]0, \infty[$. ■

Corollary

For every bounded interval $\Omega_1 \subset]0, \infty[= \Omega$, the zone control $(\Omega_1, \alpha_1 \chi_{\Omega_1})$ is strategic for (*S*).

Remark 2.11

In the case of point controls, we have a comparable result with that of Proposition 2.20.

Consider the system

$$\begin{aligned} \dot{z} &= Az + \sum_{i=1}^{p} \delta_{b_i} u_i \\ z(0) &= 0 \end{aligned} \tag{2.23}$$

where $Az = \Delta z$, $z \in \mathscr{D}(A) = \{z \in L^2(\mathbb{R}^n) : \Delta z \in L^2(\mathbb{R}^n)\}$, $b_i \in \mathbb{R}^n$, $i = 1, 2, \ldots, p$. (This equation makes sense in $D(A)'$.) We have the following.

Proposition 2.22

(1) If $n = 1$, the suite of point controls $(b_i, \delta_{b_i})_{1 \leqslant i \leqslant p}$ is strategic for equation (2.23) if and only if $p \geqslant 2$.
(2) If $n \geqslant 2$, there does not exist a finite suite of point controls which are strategic for equation (2.23).

Proof Controllability of equation (2.23) is equivalent to

$$\forall z \in L^2(\mathbb{R}^n), \qquad (S(t)z)(b_i) = 0 \qquad t \in]0, T[, \quad i = 1, \ldots, p \Rightarrow z = 0$$

By analyticity, this is equivalent to

$$\forall z \in L^2(\mathbb{R}^n), \quad \forall n \in \mathbb{N} \qquad \frac{\mathrm{d}^n}{\mathrm{d}t^n} (S(t)z)(b_i)|_{t=t_0} = 0 \qquad \forall i = 1, \ldots, p$$

and $t_0 \in]0, T[$ implies that $z = 0$.

However,

$$\frac{d^n}{dt^n}(S(t)z)(b_i)|_{t=to} = (-1)^n \mathscr{F}^{-1}(e^{-t_0\|y\|^2} \| y \|^{2n} \mathscr{F} z(y))(b_i)$$

Explicitly, for $n = 1$, we have

$$\frac{d^n}{dt^n}(S(t)z)(b_i)|_{t=to} = \frac{(-1)^n}{2\pi} \int_{\mathbb{R}} e^{i\xi b_i} \xi^{2n} e^{-t_0\xi^2} \mathscr{F} z(\xi)\, d\xi$$

$$= \frac{(-1)^n}{2\pi} \int_0^\infty \xi^{2n} e^{-t_0\xi^2} \{e^{i\xi b_i} \mathscr{F} z(\xi) + e^{-i\xi b_i} \mathscr{F} z(-\xi)\}\, d\xi$$

Thus, controllability is equivalent to

$$e^{i\xi b_i} \mathscr{F} z(\xi) + e^{-i\xi b_i} \mathscr{F} z(-\xi) = 0 \qquad \text{almost everywhere in }]0, \infty[$$

implies that $z = 0$. Then, as in Proposition 2.20, we are able to conclude (1).

For $n \geqslant 2$,

$$\frac{d^n}{dt^n}(S(t)z)(b_i)|_{t=to} = \frac{(-1)^n}{2\pi} \int_{\mathbb{R}^n} e^{i\langle \xi, b_i \rangle} \| \xi \|^{2n} e^{-t_0\|\xi\|^2} \mathscr{F} z(\xi)\, d\xi$$

and the conclusion again follows since $\| \xi \|^{2n} e^{-t_0\|\xi\|^2}$ are not dense in $L^2(\mathbb{R}^n_+)$. ■

2.5 APPLICATION TO THE LOCALIZATION OF STRATEGIC SENSORS

2.5.1 Duality

As in finite dimensions, there is a duality between observation and control. For the system

$$(S) \qquad \begin{cases} \dot{z} = Az + Bu \\ z(0) = 0 \end{cases}$$

with the operator H, defined by

$$H: L^2(0, T; \mathbb{R}^n) \to \mathscr{E}$$

$$Hu = \int_0^T S(T-\tau)Bu(\tau)\, d\tau$$

Controllability of (S) requires the determination of u such that $Hu = z_d$, $z_d \in \mathscr{E}$. Consider now the case where $\mathscr{E}$ is a Hilbert space; then

$$\langle Hu, z \rangle_{\mathscr{E}} = \left\langle \int_0^T S(T-\tau)Bu(\tau)\, d\tau, z \right\rangle_{\mathscr{E}}$$

$$= \int_0^T \langle u(\tau), B^* A^*(T-\tau)z \rangle\, d\tau$$

$$= \langle u, H^* z \rangle_{L^2(0,T;\mathbb{R}^p)}$$

Thus the adjoint H^* may be interpreted as an output operator. Given that $z \in \mathscr{E}$,

$$H^* z = B^* S^*(.)z \in L^2(0, T; \mathbb{R}^n)$$

Definition 2.1

We say that the control system

$$(S) \qquad \begin{cases} \dot{z} = Az + Bu \\ z(0) = 0 \end{cases}$$

and the observation system

$$(S_d) \qquad \begin{cases} \dot{z} = \tilde{A}z \\ y = Cz \end{cases}$$

are dual if $\tilde{A} = A^*$ and $B^* = C$.

This notion enables us to transform results on controllability into results on observability and vice versa.

Proposition 2.23

If the two systems (S) and (S_d) are dual, a necessary and sufficient condition for (S) to be weakly controllable is that (S_d) is weakly observable.

Proof (S) is weakly controllable if and only if

$$B^*S^*(t)z = 0 \qquad \forall t \in]0, T[\Rightarrow z = 0$$

and this is just the condition for weak observability of the dual system (S_d). ■

As a result of this proposition, we are able to obtain observability results immediately from the controllability results that we have already established.

Suppose, then, that we observe the system through the intermediary of q sensors $(D_i, f_i)_{1 \leqslant i \leqslant q}$ and that the system (S) is augmented by the output equation

$$(E) \qquad y(t) = \begin{bmatrix} y_1(t) \\ \vdots \\ y_q(t) \end{bmatrix} = \begin{bmatrix} \langle f_1, z(t) \rangle_{L^2(D_1)} \\ \vdots \\ \langle f_q, z(t) \rangle_{L^2(D_q)} \end{bmatrix}$$

Again we assume that (ϕ_{nj}) are eigenfunctions of A and that (λ_n) are the eigenvalues with multiplicity r_n. We have the following.

Proposition 2.24

The suite of sensors $(D_i, f_i)_{1 \leqslant i \leqslant q}$ is strategic for the system (S) augmented with the output equation (E) if and only if

(1) $q \geqslant \sup_n (r_n)$;

(2) $\operatorname{rank}(G_n) = r_n$ for all n where $(G_n)_{ij} = \langle f_i, \phi_{nj} \rangle_{L^2(D_i)}$, $i = 1, \ldots, q$, $j = 1, \ldots, r_n$.

The number of sensors necesary to have weak observability must be equal to or greater than the largest order of multiplicity of an eigenvalue of A.

2.5.2 Case of a one-dimensional domain

Consider the system (S) with $\Omega =]0, a[$ and $Az = \Delta z$, $z \in D(A) = H_0^1(\Omega) \cap H^2(\Omega)$. The eigenfunctions are given by equation (2.1) and the eigenvalues by equation (2.2). Since $r_n = 1$ for all n, one sensor may be sufficient for weak observability.

Zone sensors

Suppose that

$$D_1 = [x_1 - l_1, x_1 + l_1] \subset]0, a[, \qquad f_1 \in L^2(D_1)$$

Proposition 2.25

If f is symmetric about the line $x = x_1$, then the sensor (D_1, f_1) is not strategic for the system (S) with output equation (E) if $x_1/a \in \mathbb{Q}$.

Point sensors

In the case of point sensors, we have to be careful that the output is well defined in an analogous way to that in which we considered point controls. If a sensor is placed at $x_1 \in]0, a[$, we have the following.

Proposition 2.26

The point sensor (x_1, δ_{x_1}) is strategic for the system (S) with output (E) if and only if $x_1/a \notin \mathbb{Q}$.

Remark 2.12

(1) The above results indicate bad locations of the sensor if one wants to observe the state of a system.
(2) If the measurement is made via a zone D_1, then it is natural to assume that $f_1 = \alpha\chi_{D_1}$ and we then have the symmetry of Proposition 2.25.

2.5.3 Case of a two-dimensional domain with $\Omega =]0, \alpha_1[\times]0, \alpha_2[$

Again we assume that $Az = \Delta z$, $z \in D(A) = H_0^1(\Omega) \cap H^2(\Omega)$ and $\alpha_1^2/\alpha_2^2 \notin \mathbb{Q}$. The eigenfunctions and eigenvalues are given by equations (2.6) and (2.7) and, since $r_n = 1$ for all n, one sensor may be sufficient for weak observability. In the case of zones, two types of support are considered.

Proposition 2.27

Suppose that $a = (a_1, a_2)$ and $D_1 = \prod_{i=1}^{2} [a_i - l_i, a_i + l_i]$ or $D_1 = D(a, r)$ with $D_1 \subset \Omega$, then the zone sensor (D_1, f_1), $f_1 \in L^2(D_1)$, is not strategic for the system (S) with output (E) if for some i_0, $1 \leqslant i_0 \leqslant 2$, $a_{i0}/\alpha_{i_0} \in \mathbb{Q}$ and f_1 is symmetric about the line $x_{i_0} = a_{i0}$.

Proposition 2.28

Suppose that $a = (a_1, a_2) \in \Omega$. The point sensor (a, δ_a) is strategic for the system (S) with output (E) if and only if for all i, $1 \leqslant i \leqslant 2$, $a_i/\alpha_i \notin \mathbb{Q}$.

2.5.4 Case of a two-dimensional domain with $\Omega = D(0, 1)$

We have seen that, in this case, $\sup_n(r_n) = 2$ and so it is necessary to have at least two sensors for the observability problem.

Proposition 2.29 Zone sensors

Suppose that the supports D_1 and D_2 are symmetric about $\theta = \theta_1$ and $\theta = \theta_2$, then the pair of sensors $(D_i, f_i)_{1 \leqslant i \leqslant 2}$ are not strategic for the system (S) with output (E) if, for all i, $1 \leqslant i \leqslant 2, f_i$ is symmetric about $\theta = \theta_i$ and $(\theta_1 - \theta_2)/\pi \in \mathbb{Q}$.

Proposition 2.30 Point sensors

If the sensors have polar coordinates $c_i(\rho_i, \theta_i)$, $1 \leqslant i \leqslant 2$, then

(1) the pair of point sensors $(c_i, \delta_{c_i})_{1 \leqslant i \leqslant 2}$ are not strategic for the system (S) with output (E) if $(\theta_1 - \theta_2)/\pi \in \mathbb{Q}$;
(2) if $\rho_1 = \rho_2$, the pair of point sensors (c_i, δ_{c_i}) are strategic for the system (S) with output (E) if and only if $(\theta_1 - \theta_2)/\pi \notin \mathbb{Q}$.

Remark 2.13

It is possible to envisage sensors of the filament type. Consider the case where

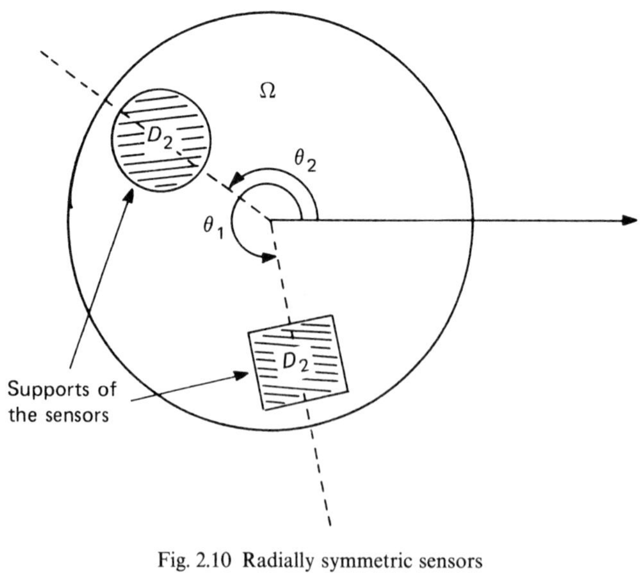

Fig. 2.10 Radially symmetric sensors

$\Omega =]0, \alpha_1[\times]0, \alpha_2[$ and suppose that the observation on the curve

$$\Gamma = \mathrm{Im}\,(\gamma)$$

with $\gamma \in C^1(0, 1)$, then we have the following.

Corollary

The filament sensor (Γ, δ_Γ) is not strategic for the system (S) augmented with the output (E) if $x_0/\alpha_1 \in \mathbb{Q}$ and Γ is symmetric about the line $x_1 = x_0$.

2.5.5 Sensors on the boundary

Suppose that we consider the system

$$\frac{\partial z}{\partial t} - \Delta z = 0 \qquad \text{in } \Omega \times]0, T[$$

with Neumann control

$$\frac{\partial z(\xi, t)}{\partial v} = \sum_{i=1}^{q} f_i(\xi) u_i(t) \qquad \xi \in \partial\Omega, \quad t \in]0, T[$$

then the dual observation system is

$$\begin{aligned} &\frac{\partial z}{\partial t} - \Delta z = 0 && \text{in } \Omega \times]0, T[\\ &z(x, 0) = z_0(x) && \text{in } \Omega \\ &\frac{\partial z(\xi, t)}{\partial v} = 0 && \text{in } \partial\Omega \times [0, T[\end{aligned} \tag{2.24}$$

with output

$$(E) \qquad y(t) = \begin{bmatrix} \langle f_1, z(t) \rangle_{L^2(\Gamma_1)} \\ \vdots \\ \langle f_q, z(t) \rangle_{L^2(\Gamma_q)} \end{bmatrix}$$

$(\Gamma_i, f_i)_{1 \leqslant i \leqslant q}$ is a suite of q sensors on the boundary with $\Gamma_i \subset \partial\Omega$ and $f_i \in L^2(\Gamma_i)$.

Similarly, if we consider instead the Dirichlet control problem with

$$z(\xi, t) = \sum_{i=1}^{q} f_i(\xi) u_i(t) \qquad \xi \in \partial\Omega, \quad t \in]0, T[$$

the dual observation problem is

$$\begin{aligned} &\frac{\partial z}{\partial t} - \Delta z = 0 && \text{in } \Omega \times]0, T[\\ &z(x, 0) = z_0(x) && \text{in } \Omega \\ &z(\xi, t) = 0 && \text{in } \partial\Omega \times]0, T[\end{aligned} \tag{2.25}$$

with output

$$(E)' \qquad y(t) = \begin{bmatrix} \left\langle f_1 \frac{\partial z}{\partial v} \right\rangle_{L2(\Gamma_1)} \\ \vdots \\ \left\langle f_q, \frac{\partial z}{\partial v} \right\rangle_{L^2(\Gamma_q)} \end{bmatrix}$$

If (ϕ_{nj}) are the eigenfunctions of Δ for the Dirichlet problem (and ψ_{nj} those for the Neumann problem), we have the following characterization.

Proposition 2.31

The suite of sensors on the boundary $(\Gamma_i, f_i)_{1 \leqslant i \leqslant q}$ is strategic for the system (2.24) (or the system (2.25)) with output (E) (or $(E)'$) if and only if

(1) $q \geqslant \sup_n (r_n)$;

(2) $\operatorname{rank}(G_n) = r_n$ for all n with $(G_n)_{ij} = \langle f_i, \psi_{nj} \rangle_{L^2(\Gamma_i)}$ for the system (2.24) and $(G_n)_{ij} = \langle f_i, (\partial\phi_{nj}/\partial v) \rangle_{L^2(\Gamma_i)}$ for the system (2.25), $i = 1, \ldots, q$, $j = 1, \ldots, r_n$.

Application to a two-dimensional domain with $\Omega = \prod_{i=1}^{2}]0, \alpha_i[$

Consider the case $\Omega = \prod_{i=1}^{2}]0, \alpha_i[$ with $\alpha_1^2/\alpha_2^2 \notin \mathbb{Q}$, and one sensor is located in the same place as the control in Fig. 2.6; then, if f_1 is symmetric about the line $x_1 = a_1$, the sensor (Γ_1, f_1) is not strategic for the system (2.24) if there exists an integer m such that $2a_1 m/\alpha_1$ is odd. For the system (2.25), the sensor (Γ_1, f_1) is not strategic if f_1 is symmetric about $x_1 = a_1$ and $a_1/\alpha_1 \in \mathbb{Q}$.

Application to a circular domain with $\Omega = D(0, 1)$

In the case where $\Omega = D(0, 1)$, it is necessary to have at least two sensors for weak observability. The results are identical with those in Remark 2.6 concerning boundary controls.

Remark 2.14

The case of point sensors on the boundary can be treated in a similar manner. The characterization is the same as that given in Proposition 2.31 but it is necessary to take $(G_n)_{ij} = \psi_{nj}(b_i)$ for the system (2.24) and $(G_n)_{ij} = (\partial\phi_{nj}(b_i)/\partial v)$ for the system (2.25). The results are the same as for point controls on the boundary.

2.6 CHOICE OF CONTROLS AND SENSORS FOR STABILIZABILITY AND DETECTABILITY

Consider the system

$$(S) \qquad \begin{cases} \dot{z} = Az + Bu \\ z(0) = z_0 \end{cases}$$

with output equation

$$(E) \qquad y = Cz$$

Proposition 1.6 and 1.7 give a characterization of stabilizability by a choice of controls and detectability by a choice of sensors. We now apply these results to a number of problems with different geometries. Since we do not want to assume that the system is already stable, we let

$$Az = \Delta z + \alpha z$$

$z \in H_0^1(\Omega) \cap H^2(\Omega)$. So for the case of zones the system is

$$\begin{aligned}
&\frac{\partial z}{\partial t} = \Delta z + \alpha z + \sum_{i=1}^{p} g_i u_i && \text{in } \Omega \times]0, T[\\
&z(x, 0) = z_0(x) && \text{in } \Omega \\
&z(\xi, t) = 0 && \text{in } \partial\Omega \times]0, T[
\end{aligned} \tag{2.26}$$

where $\Omega_i \subset \Omega$, $g_i \in L^2(\Omega_i)$, $i = 1, \ldots, p$. The output equation is

$$(E) \qquad y = \begin{bmatrix} y_1 \\ \vdots \\ y_q \end{bmatrix} = \begin{bmatrix} \langle f_1, z \rangle_{L^2(D_1)} \\ \vdots \\ \langle f_q, z \rangle_{L^2(D_q)} \end{bmatrix}$$

where $D_i \subset \Omega$, $f_i \in L^2(D_i)$, $i = 1, \ldots, q$.

2.6.1 Case of a one-dimensional domain

We consider the case where $\Omega =]0, a[$ and let $p - q = 1$. If Ω_1 (or D_1) equals $[x_1 - l_1, x_1 + l_1] \subset \Omega$, then we have the following.

Proposition 2.32

If g_1 (or f_1) is symmetric about $x = x_1$, then the system is stabilizable (or detectable if, for some i, $1 \leqslant i \leqslant J$, $ix/a \subset \mathbb{N}$.

In the case of a single point control (or sensor) placed at $x_1 \subset]0, a[$, we have the following.

Proposition 2.33

The system is stabilizable (or detectable) if and only if $ix_1/a \notin \mathbb{N}$ for all $i = 1, \ldots, J$.

2.6.2 Case of a two-dimensional domain with $\Omega =]0, \alpha_1[\times]0, \alpha_2[$

We assume that $\alpha_1^2/\alpha_2^2 \notin \mathbb{Q}$ so that $r_n = 1$ for all n and consider only the case $p = q = 1$.

Proposition 2.34 Case of zones

Suppose that $a = (a_1, a_2)$ and Ω_1 (or D_1) is given by $\Omega_1 = \prod_{i=1}^{2} [a_i - l_i, a_i + l_i]$ or $\Omega_1 = D(a, r)$, $\Omega_1 \subset \Omega$. Then the system is not stabilizable (or detectable) if for i_0,

$1 \leqslant i_0 \leqslant 2$, $ia_{i_0}/\alpha_{i_0} \in \mathbb{N}$ for some i, $1 \leqslant i \leqslant J$, and g_1 (or f_1) is symmetric about the line $x_{i_0} = a_{i_0}$.

Proposition 2.35 Case of points

If $a = (a_1, a_2) \in \Omega$, then the control (or sensor) (a, δ_a) can stabilize (or detect) the system if and only if for all j, $1 \leqslant j \leqslant 2$, $ia_j/\alpha_j \notin \mathbb{N}$ for all i, $1 \leqslant i \leqslant J$.

Remark 2.15

In the case of controls (or sensors) of the filament type with symmetry about $x_1 = x_0$, the system is not stabilizable (or detectable) if $ix_0/\alpha_1 \in \mathbb{N}$ for some i, $1 \leqslant i \leqslant J$.

2.6.3 Case of a two-dimensional domain with $\Omega = D(0, 1)$

We consider the system (2.26) so that the eigenvalues of A are

$$\lambda_{nm} = \alpha^2 - \beta_{nm}^2$$

where β_{nm} are the roots of the Bessel function J_n. As before, we assume that $\lambda_{mn} > 0$ for $n = 0, \ldots, J$ and $m = 1, \ldots, K$. The order of multiplicity is 2 if $n \neq 0$ and 1 if $n = 0$. It is necessary therefore to consider at least two controls (or sensors) in order to stabilize (or detect) the system. The system is stabilizable if

$$\langle g_1, \phi_{0m} \rangle_{L^2(\Omega_1)}^2 + \langle g_2, \phi_{0m} \rangle_{L^2(\Omega_2)}^2 \neq 0$$

for $m = 1, \ldots, K$ and

$$\det \begin{bmatrix} \langle g_1, \phi_{nm_1} \rangle_{L^2(\Omega_1)} & \langle g_1, \phi_{nm_2} \rangle_{L^2(\Omega_1)} \\ \langle g_2, \phi_{nm_1} \rangle_{L^2(\Omega_2)} & \langle g_2, \phi_{nm_2} \rangle_{L^2(\Omega_2)} \end{bmatrix} \neq 0$$

for $n = 1, \ldots, J$, $m = 1, \ldots, K$, where ϕ_{nm} are the eigenfunctions given in equation (2.10). A similar characterization can be given for detectability.

We then have the following results.

Proposition 2.36 Case of zones

Suppose that, for all i, $1 \leqslant i \leqslant 2$, Ω_i (or D_i) is symmetric about $\theta = \theta_i$. Then the system is not stabilizable (or detectable) if g_i (or f_i) is symmetric about $\theta = \theta_i$ and $i(\theta_1 - \theta_2)/\pi \in \mathbb{N}$ for some i, $1 \leqslant i \leqslant J$.

Proposition 2.37 Case of points

Suppose that the controls (or sensors) are located at $c_i = (\rho_i, \theta_i) \in \Omega$, $i = 1, 2$. Then

(1) the system is not stabilizable (or detectable) if $i(\theta_1 - \theta_2)/\pi \in \mathbb{N}$ for some i, $1 \leqslant i \leqslant J$;
(2) if $\rho_1 = \rho_2$, the system is stabilizable (respectively detectable) if and only if $i(\theta_1 - \theta_2)/\pi \notin \mathbb{N}$ for all $i = 1, \ldots, J$.

2.6.4 Boundary case

We consider the case of controls and sensors located on the boundary $\partial\Omega$ of Ω. For example, the system with Dirichlet control is

$$\begin{aligned} \frac{\partial z}{\partial t} &= \Delta z + \alpha z && \text{in } \Omega \times]0, T[\\ z(x,0) &= z_0(x) && \text{in } \Omega \\ z(\xi, t) &= \sum_{i=1}^{p} g_i(\xi)u_i(t) && \text{in } \partial\Omega \times]0, T[\end{aligned} \tag{2.27}$$

and the output equation

$$(E) \qquad y = \begin{bmatrix} y_1 \\ \vdots \\ y_q \end{bmatrix} = \begin{bmatrix} \left\langle f_1, \frac{\partial z}{\partial v} \right\rangle_{L^2(D_1)} \\ \vdots \\ \left\langle f_q, \frac{\partial z}{\partial v} \right\rangle_{L^2(D_q)} \end{bmatrix}$$

$(\Gamma_i, g_i)_{1 \leqslant i \leqslant p}$ (or $(D_i, f_i)_{1 \leqslant i \leqslant q}$) are the controls (or sensors) located at the boundary. We always use (ϕ_{nj}) (or (ψ_{nj})) to denote the eigenfunctions for the Dirichlet (or Neumann) problems.

Proposition 2.38 Stabilizability with boundary control

The system excited by the suite of boundary controls $(\Gamma_i, g_i)_{1 \leqslant i \leqslant p}$ is stabilizable if and only if

(1) $p \geqslant \sup_{1 \leqslant n \leqslant J} (r_n)$;
(2) $\text{rank}\,(G_n) = r_n$ for $n = 1, \ldots, J$ where $(G_n)_{ij} = \langle g_i, (\partial\phi_{nj}/\partial v)\rangle_{L^2(\Gamma_i)}$ for the Dirichlet case and $(G_n)_{ij} = \langle g_i, \psi_{nj}\rangle_{L^2(\Gamma_i)}$ for the Neumann case, $i = 1, \ldots, p$, $j = 1, \ldots, r_n$,

Proposition 2.39 Detectability with boundary sensors

If the observations are made through q sensors $(D_i, f_i)_{1 \leqslant i \leqslant q}$ on the boundary, then the system with Dirichlet boundary conditions and output (E) is detectable if and only if

(1) $q \geqslant \sup_{1 \leqslant n \leqslant J} (r_n)$;
(2) $\text{rank}\,(G_n) = r_n$ for all $n = 1, \ldots, J$ with $(G_n)_{ij} = \langle f_i, (\partial\phi_{nj}/\partial v)\rangle_{L^2(D_i)}$, $i = 1, \ldots, q$, $j = 1, \ldots, r_n$.

A similar result holds for Neumann boundary conditions and

$$y = \begin{bmatrix} y_1 \\ \vdots \\ y_q \end{bmatrix} = \begin{bmatrix} \langle f_1, z \rangle_{L^2(D_1)} \\ \vdots \\ \langle f_q, z \rangle_{L^2(D_q)} \end{bmatrix}$$

with $(G_n)_{ij}$ replaced by $\langle f_i, \psi_{nj}\rangle_{L^2(D_i)}$, $i = 1, \ldots, q$, $j = 1, \ldots, r_n$.

Application to a two-dimensional domain with $\Omega = \prod_{i=1}^{2}]0, \alpha_i[$

We always suppose that $\alpha_1^2/\alpha_2^2 \notin \mathbb{Q}$ and suppose that the controls are those given in Figs 2.6 and 2.7.

Proposition 2.40 Support on one side

If g_1 (or f_1) is symmetric about $x_1 = a_1$, then

(1) in the case of Dirichlet boundary conditions the system is not stabilizable (or detectable) if $ia_1/\alpha_1 \in \mathbb{N}$ for some $i = 1, \ldots, J$;
(2) in the case of Neumann boundary conditions the system is not stabilizable (or detectable) if there exists $i, 1 \leqslant i \leqslant J$, such that $2ia_1/\alpha_1$ is odd.

Proposition 2.41 Support on two sides

If g_1 (or f_1) is doubly symmetric about the lines $x_1 = a_1$ and $x_2 = a_2$, then

(1) in the case of Dirichlet boundary conditions the system is not stabilizable (or detectable) if $ia_1/\alpha_1 \in \mathbb{N}$ for some i, $1 \leqslant i \leqslant J$ and $ia_2/\alpha_2 \in \mathbb{N}$ for some i, $1 \leqslant i \leqslant J$;
(2) in the case of Neumann boundary conditions the system is not stabilizable (or detectable) if there exist two integers m, n, $1 \leqslant m, n \leqslant J$ such that $2a_1 n/\alpha_1$ and $2a_2 m/\alpha_2$ are odd.

Application to a circular domain with $\Omega = D(0, 1)$

It is necessary to have at least two controls (or sensors) and if g_i (or f_i) are symmetric about $\theta = \theta_i$, $i = 1, 2$, then the system is not stabilizable (or detectable) if $i\,(\theta_1 - \theta_2)/\pi \in \mathbb{N}$ for some i, $1 \leqslant i \leqslant J$.

Remark 2.16

In the case of point controls and sensors, similar results can be obtained. Suppose for example that we consider $\Omega = D(0, 1)$ with controls (or sensors) at polar coordinates $(1, \theta_i)$, $i = 1, 2$, then the system is not stabilizable (or detectable) if $i(\theta_1 - \theta_2)\pi \in \mathbb{N}$ for some i, $1 \leqslant i \leqslant J$.

Chapter 3
Optimization of the parameters of a control

If one considers a system excited by p controls it is natural to associate with the system a cost function. The object is to choose the controls to minimize the cost function. However, this will depend on how we assign the following:

(1) the location of the controls;
(2) the spatial distributions of the controls on their supports;
(3) the number of controls.

This type of problem must be resolved at a conceptual level, since in general it is not reasonable, or even impossible, to act on all the parameters which define the controls. For example, in a heating system, we may have choice as to where the burners are placed or whether or not some of the zones of the burner are activated, but we shall not be able to redesign the burner completely.

In this chapter, we shall be interested in the choice of the parameters for zone controls (Ω_i, g_i). We shall examine first the problem of optimizing the support Ω_i. For this, one must choose the distribution g_i. We shall give some results for one-dimensional systems. For the general case a direct approach is taken where g_i is introduced as a new control variable. As far as the number of controls is concerned, we do not develop any theory but make some comments in the light of our experience in numerical simulations.

3.1 OPTIMAL LOCATION OF CONTROLS

3.1.1 Choice of criterion

In order to resolve the optimal location problem, we consider three stages.

(1) The choice of a criterion depends on the parameters of the support of the controls. Different criteria are proposed in the literature and all the developments which

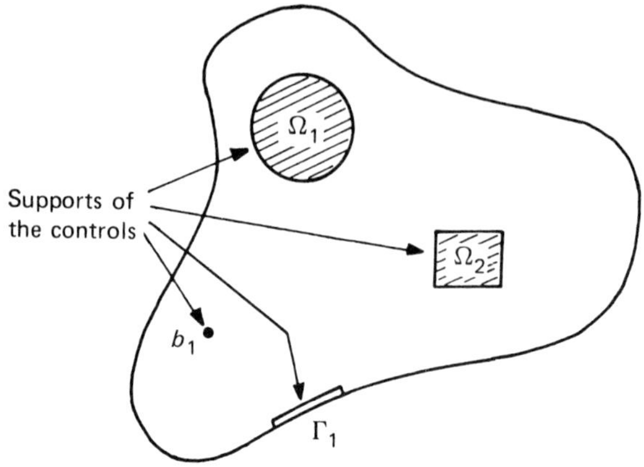

Fig. 3.1

follow may be applied equally well to these other criteria. We shall choose an energy-type criterion with a penalty on the final state; this has the advantage of giving an explicit formula for the minimum cost.

(2) The criterion should be minimized. Direct methods can be used where the cost function is expressed in terms of the parameters of the support. A more elegant approach is by using the mathematical tools of optimum design. These have been used for determining the geometry of a diffuser of heat in a satellite or electromagnet. This second approach has the advantage of being applicable to arbitrary supports.

(3) The choice of a numerical method leads to the calculation of the various parameters. For this, it is necessary to have efficient algorithms well adapted to the problem. These algorithms must essentially have good performance, since at each iteration an optimal control problem must be solved and a minimum cost calculated, and this involves a discretized calculation of the gradient. The calculation of the gradient of an optimal cost as a function of the parameters of the support is often very complex and requires a large number of computations.

The systems that we consider are

$$(S) \qquad \begin{cases} \dot{z} = Az + Bu & 0 < t < T \\ z(0) = z_0 \end{cases}$$

with the following hypotheses.

(1) The state space $\mathscr{E}$ is a separable Hilbert space (we take $\mathscr{E} = L^2(\Omega)$).
(2) A is self-adjoint with compact resolvent and generates a strongly continuous semigroup $(S(t))_{t \geqslant 0}$ on $\mathscr{E}$.
(3) $u = (u_1, \ldots, u_p) \in L^2(0, T; \mathbb{R}^p)$.
(4) $B \in \mathscr{L}(\mathbb{R}^p, \mathscr{E})$.

The choice of criterion is an important step. If the criterion is a general quadratic one, then the resolution of the optimal control problem requires the solution of a dynamic Riccati equation. Moreover, this must be carried out at each iteration as we search for the optimal parameters of the supports of the controls. We do not examine this criterion.

A criterion which is more simple to exploit is

$$J = \min_{u \in \mathscr{U}_{\text{ad}}} (\| u \|^2_{L^2(0,T;\mathbb{R}^p)}) \tag{3.1}$$

where $\mathscr{U}_{\text{ad}}$ is the set of controls which steer z_0 (at $t = 0$) to a z_d (at $t = T$) in the space $\mathscr{E}$. The optimal control is an open loop of the form

$$u(t) = F(z_0, z_d, t) \tag{3.2}$$

We shall develop, in this chapter, two approaches to show how it is possible to determine numerically the zones of action which minimize J. Clearly, we shall rely heavily on the ideas of controllability introduced in Chapter 2. Also the choice of the criterion (3.1) is quite reasonable if the objective is to drive the system from an initial state to some desired profile.

Suppose that (S) is excited by p zone controls (Ω_i, g_i), $\Omega_i \subset \Omega$ and $g_i \in L^2(\Omega_i)$. We take $z_0 = 0$ and consider again the operator H defined by

$$H: u \in L^2(0, T; \mathbb{R}^p) \to Hu \in \mathscr{E}$$

with

$$Hu = \int_0^T S(T - \tau) Bu(t)\, \mathrm{d}t \tag{3.3}$$

We have seen that

$$H^*: \mathscr{E} \to L^2(0, T; \mathbb{R}^p)$$

is given by

$$H^* z(.) = B^* S^*(T - .) z \tag{3.4}$$

Let $M = HH^*$, then we have the following.

Lemma 3.1

The suite of controls $(\Omega_i, g_i)_{1 \leqslant i \leqslant p}$ is strategic for the system (S) if and only if M is positive.

This follows immediately from Definition 1.3 and Proposition 1.2. ■

If $z_d \in \mathscr{E}$, the solution of the minimization problem (3.1) is given as follows.

Theorem 3.1

The suite of controls $(\Omega_i, g_i)_{1 \leqslant i \leqslant p}$ is strategic for the system (S) and, if $z_d \in \operatorname{Im}(M)$, then the unique control u^* which solves equation (3.1) is given by

$$u^* = H^\dagger z_d \tag{3.5}$$

whose $H^\dagger$ is the pseudo-inverse of H. Moreover,

$$\| u^* \|^2_{L^2(0,T;\mathbb{R}^p)} = \langle z_d, M^{-1} z_d \rangle \tag{3.6}$$

For the proof, it is necessary to recall the following.

Recall Generalized inverses

If X and Y are two Hilbert spaces and $H \in \mathscr{L}(X, Y)$, considered as a map from $\operatorname{Ker}(H)^\perp \to \operatorname{Im}(H)$, H admits an inverse H^{-1}. One defines the pseudo-inverse of H, $H^\dagger$

as a linear extension of $(H|_{\mathrm{Ker}(H)^\perp})^{-1}$, where

$$D(H^\dagger) = \mathrm{Im}\,(H) + \mathrm{Im}\,(H)^\perp$$
$$\mathrm{Ker}\,(H^\dagger) = \mathrm{Im}\,(H)^\perp$$

In the case where $\mathrm{Im}\,(H)$ is closed, then $Y = \mathrm{Im}\,(H) + \mathrm{Im}\,(H)^\perp$ and hence $H^\dagger$ is bounded.
For all y, there exists an x' such that

$$\| Hx' - y \| = \inf_{x \in X} \| Hx - y \|$$

and all such x' satisfy the normal equations

$$H^*Hx' = H^*y$$

The existence of x' is characterized by

$$P_{\overline{\mathrm{Im}(H)}}\, y \in \mathrm{Im}\,(H)$$

x' is therefore an extremal solution of $Hx = y$. Since these solutions are extremal, there exists a unique x^* with minimum norm characterized by

$$P_{\overline{\mathrm{Ker}\,(H)}}\, x^* \in \mathrm{Ker}\,(H)$$

and, if $y \in \mathrm{Im}\,(HH^*)$, one has

$$x^* = H^\dagger y = H^*(HH^*)^{-1} y$$

For more details, see Erdelyi (1972), Benchimol (1977) and Berrahmoune (1984)

Proof of Theorem 3.1 In the preceding, we took

$$X = L^2(0, T; \mathbb{R}^p), \qquad Y = \mathscr{E}$$

H is given by equation (3.3). If the suite of controls (Ω_i, g_i) is strategic, then M^{-1} exists and, for all $z_d \in \mathrm{Im}\,(M)$, the equation

$$Hu = z_d$$

has a minimum norm solution given by

$$u^* = H^\dagger z_d = H^*(HH^*)^{-1} z_d$$

Explicitly, we have

$$u^*(t) = B^*S^*(T-t)\left\{\int_0^T S(T-\tau)BB^*S^*(T-\tau)\,d\tau\right\}^{-1} z_d \tag{3.7}$$

Equation (3.6) follows immediately. ■

Remark 3.1

In the case where $z_0 \neq 0$, one obtains the same result by replacing z_d by $\tilde{z}_d$ where

$$\tilde{z}_d = z_d - S(T)z_0 \tag{3.8}$$

and one assumes that, in this case, $\tilde{z}_d \in \mathrm{Im}\,(M)$.

3.1.2 Approximation of the criterion

Even if the criterion chosen is relatively simple and leads to an explicit control, we still have the difficulty associated with the infinite dimensional nature of the problem. So we have to make approximations.

For the class of parabolic systems, this difficulty may be easily overcome by projections onto finite dimensional subspaces. Suppose that (ϕ_n) are eigenfunctions of A with corresponding eigenvalues (λ_n) of multiplicity one. Consider then, for all $s > 0$,

$$S_N(s) \in \mathscr{L}(\mathbb{R}^N), \qquad S_N(s) = (S_{ij})_{\substack{1 \leqslant i \leqslant N \\ 1 \leqslant j \leqslant N}}$$

and

$$S_{ij}(s) = \begin{cases} e^{\lambda_i s} & \text{if } 1 \leqslant i = j \leqslant N \\ 0 & \text{if } 1 \leqslant i \neq j \leqslant N \end{cases} \tag{3.9}$$

Also

$$B_N \in \mathscr{L}(\mathbb{R}^p, \mathbb{R}^N), \qquad B_N = (B_{ij})_{\substack{1 \leqslant i \leqslant N \\ 1 \leqslant j \leqslant N}}$$

and

$$B_{ij} = \langle g_j, \phi_i \rangle_{L^2(\Omega_j)}$$

Set

$$M_N = \int_0^T S_N(T-\tau) B_N B_N^* S_N^*(T-\tau)\, \mathrm{d}\tau \tag{3.10}$$

Then $M_N \in \mathscr{L}(\mathbb{R}^N)$ and has elements

$$M_{ij} = \sum_{k=1}^{p} B_{ik} B_{jk} \frac{e^{(\lambda_i + \lambda_j)T} - 1}{\lambda_i + \lambda_j} \tag{3.11}$$

The controllability hypothesis implies the existence of

$$V_N = M_N^{-1} \in \mathscr{L}(\mathbb{R}^N) \tag{3.12}$$

Suppose the reduced system is

$$(S_N) \qquad \begin{cases} \dot{z}_N = A_N z_N + B_N u \\ z_N(0) = 0 \end{cases}$$

where $A_N = \operatorname{diag}\{\lambda_1, \lambda_2, \ldots, \lambda_N\}$. Let

$$z = \begin{bmatrix} z_N \\ z_r \end{bmatrix}$$

where the subscript r stands for the residual part of the state. If

$$z_d = \begin{bmatrix} z_{dN} \\ z_{dr} \end{bmatrix}$$

then it is easy to show the following.

Theorem 3.2

$$u_N^*(t) = B_N^* S_N^*(T-t) V_N z_{dN} \tag{3.13}$$

is the unique control of the minimum norm which realizes the minimum of

$$\| z_{dN} - z_N(T) \|_{\mathbb{R}^N}^2 \tag{3.14}$$

■

The control in equation (3.13) is a vector of dimension p, where jth component is given by

$$u^*_{Nj}(t) = \sum_{k=1}^{N} B_{kj} e^{\lambda_k(T-t)} \sum_{m=1}^{N} V_{km} z_{dm} \tag{3.15}$$

Remark 3.2

In the case where $z_N(0) \neq 0$, it is necessary to replace z_d by

$$\tilde{z}_d = \begin{bmatrix} \tilde{z}_{dN} \\ \tilde{z}_{dr} \end{bmatrix}$$

where $\tilde{z}_{dN} \in \mathbb{R}^N$ (see Remark 3.1) and then

$$\| u^*_N \|^2 = \sum_{i,j=1}^{N} V_{ij} \tilde{z}^2{}_{di} \tilde{z}^2{}_{dj} \tag{3.16}$$

3.1.3 Direct approach

The direct approach consists of considering a criterion given by equation (3.16) of minimum energy type as a real function f of parameters $\alpha_1, \ldots, \alpha_n$ which define the support of the control:

$$f(\alpha_1, \ldots, \alpha_n) = \| u^*_N \|^2 = \min \| u_N \|_{L^2(0,T;\mathbb{R}^p)}$$

In the expression $\| u^*_N \|^2$ the parameters α_i appear in the matrices B_N and V_N. A theoretical development based on this approach is not very useful. Instead, we utilize a direct approach based on a calculation of the gradient.

Denote by

$$\Omega(\alpha_1, \alpha_2, \ldots, \alpha_n)$$

the support of a control. Then the flow chart of the algorithm is represented in Fig. 3.2. The minimization stage is carried out by algorithms in the Modulopt and the Harwell package. The resolution of the optimal control problem is usually very long for distributed parameter systems but, since the emphasis in our work is on the choice of the structure of the control, we consider only simple situations (one dimensional with a diagonal system). Then we may use the development in section 3.1.2. In section 3.1.4, we shall look at a two-dimensional domain.

Example 3.1

Consider the system

$$\begin{aligned} &\frac{\partial y(x,t)}{\partial t} = \alpha \frac{\partial^2 y(x,t)}{\partial x^2} + \sum_{i=1}^{p} g_i(x) u_i(t) \qquad x \in]0,1[\\ &y(x,0) = 0 \\ &y(0,t) = y(1,t) = 0 \end{aligned} \tag{3.17}$$

The operator $\alpha(\partial^2/\partial x^2)$ has a complete set of eigenfunctions

$$\phi_i(x) = 2^{1/2} \sin(i\pi x)$$

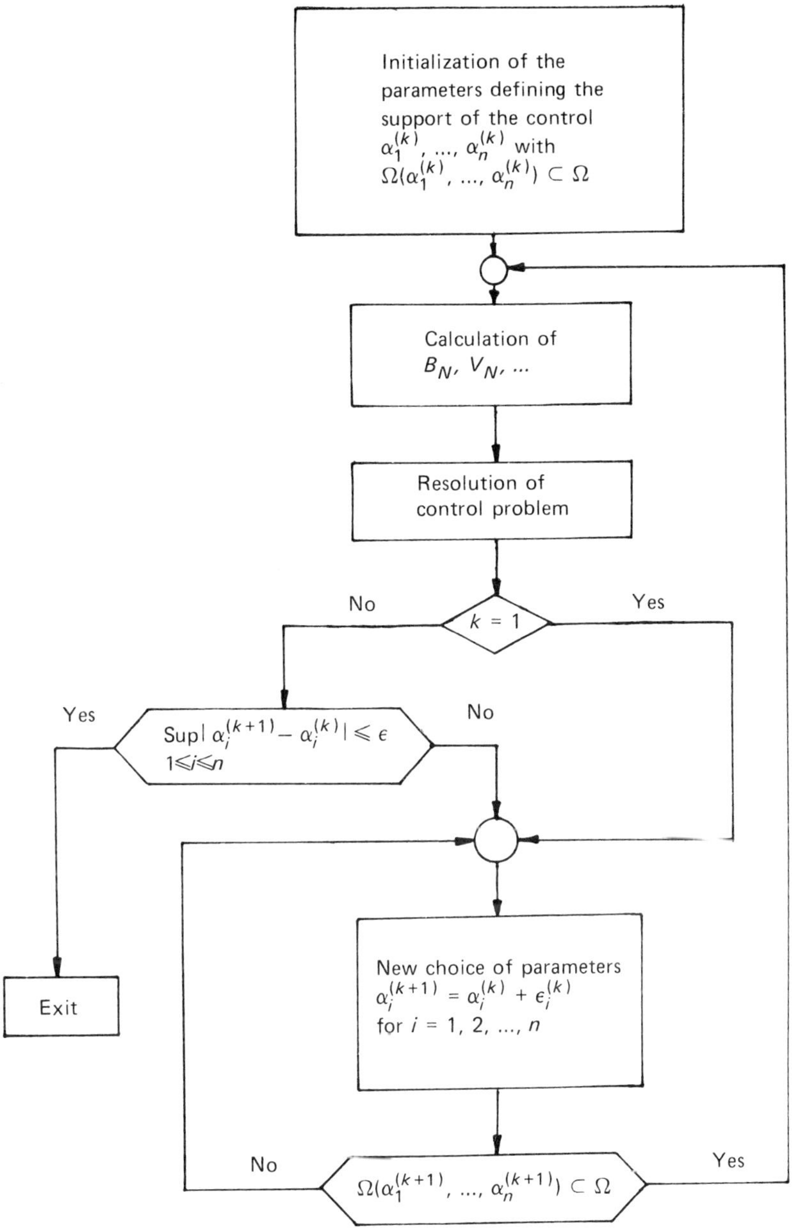

Fig. 3.2 Flow chart of the algorithm

with the eigenvalues given by

$$\lambda_i = -\alpha i^2 \pi^2$$

Suppose that the controls $(\Omega_i, g_i)_{1 \leqslant i \leqslant p}$ have

$$\Omega_i = [x_i - 0.1, x_i + 0.1]$$

Table 3.1 Optimal location of controls

Number of controls p	Optimal location of $\Omega_i = [x_i - 0.1, x_i + 0.1]$ $i = 1, 2, \ldots, p$	$\|u_N^*\|^2$	$\|z_N(T) - z_{dN}\|^2$
1	$x_1 = 0.606$	266.14	0.66×10^{-2}
2	$x_1 = 0.733$ $x_2 = 0.388$	1.63	0.79×10^{-3}
3	$x_1 = 0.779$ $x_2 = 0.347$ $x_3 = 0.623$	1.08	0.54×10^{-3}
4	$x_1 = 0.791$ $x_2 = 0.317$ $x_3 = 0.706$ $x_1 = 0.5203$	0.825	0.35×10^{-3}

and

$$g_i(x) = \begin{cases} 1 & x \in \Omega_i \\ 0 & x \notin \Omega_i \end{cases}$$

Then the only parameters to be determined are the centres x_i of the zones Ω_i, with $\Omega_i \subset]0, 1[, i = 1, \ldots, p$.

For this choice of control, we have

$$B_{ij} = \frac{2^{3/2}}{i\pi} \sin\left(\frac{i\pi}{10}\right) \sin(i\pi x_j) \tag{3.18}$$

In the numerical study, we chose

$$z_d(x) = x(1 - x^3) \qquad x \in]0, 1[$$
$$\alpha = 0.5$$
$$T = 1$$

The approximation was carried out via projections onto subspaces of dimension $N = 4$ (Table 3.1).

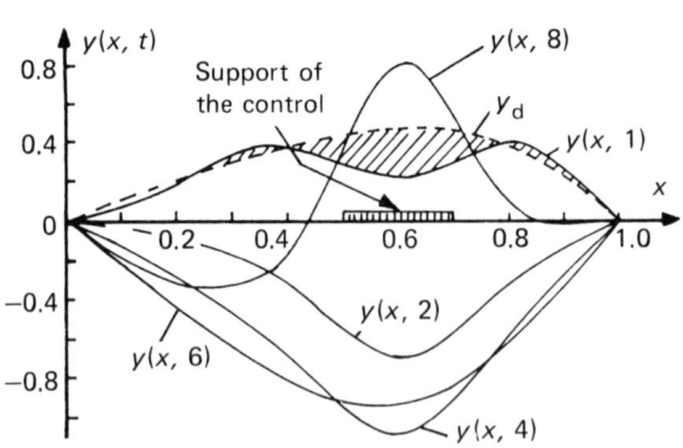

Fig. 3.3 Evolution of the state in the case $p = 1$

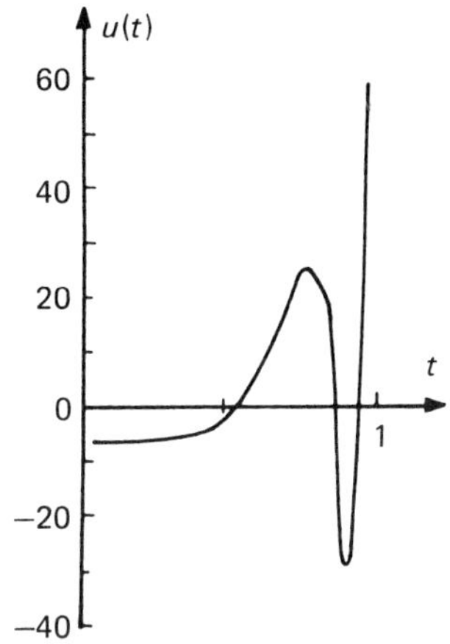

Fig. 3.4 Evolution of the control in the case $p = 1$, $x_1 = 0.606$

We also show how the various controls and states evolved in time (Figs. 3.3–3.7).

The work on these examples demonstrated the importance of determining a good location for the controls. Even though the theoretical approach by optimization techniques may not be too fruitful, it is important not to ignore these problems when designing the system.

Remark 3.3

(1) Many other examples have been treated by El Jai (1977, 1978).
(2) When one introduces the width of the zone as a parameter for optimization, many different numerical studies have shown that the optimal zones have the largest width.
(3) As shown in Table 3.1, the energy and the final error decrease rapidly as $1/p$. All simulations with various criteria confirm this fact. An attempt to demonstrate this dependence has been given in by El Jai (1977). In the particular case where one

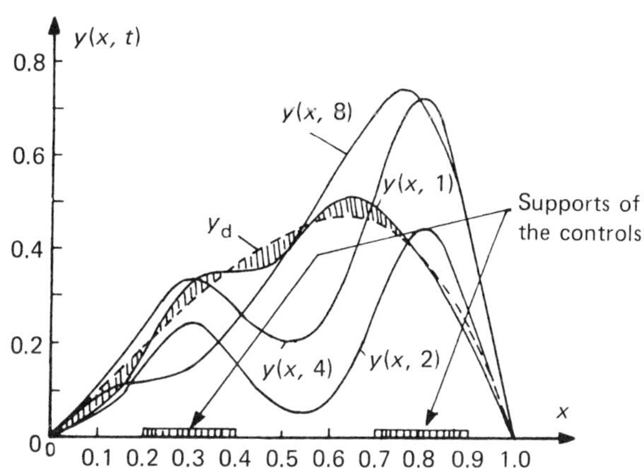

Fig. 3.5 Evolution of the state for $p = 2$ localized arbitrarily

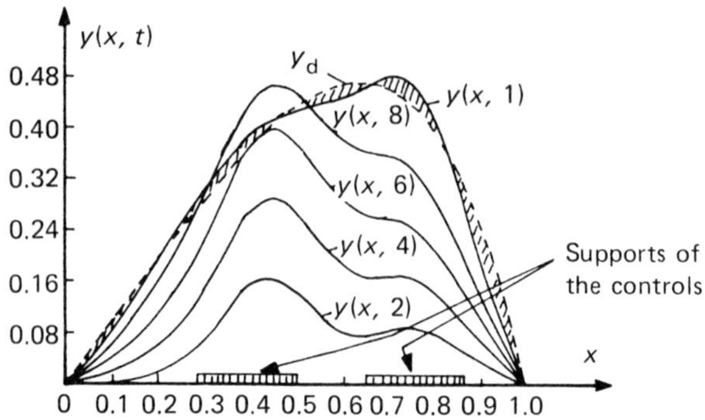

Fig. 3.6 Evolution of state for $p = 2$ localized optimally

projects the system onto a subspace of small dimension, it is easy to see that the cost is inversely proportional to

$$\sum_j \sum_{i=1}^{p} B_{ij}^2$$

For the optimal choice of p an adequate criterion is not clear.

(4) In all the studies relating to the supports of the controls (with different geometries, Ω_i connected) the minimum of a quadratic criterion decreases as Ω_i is augmented.

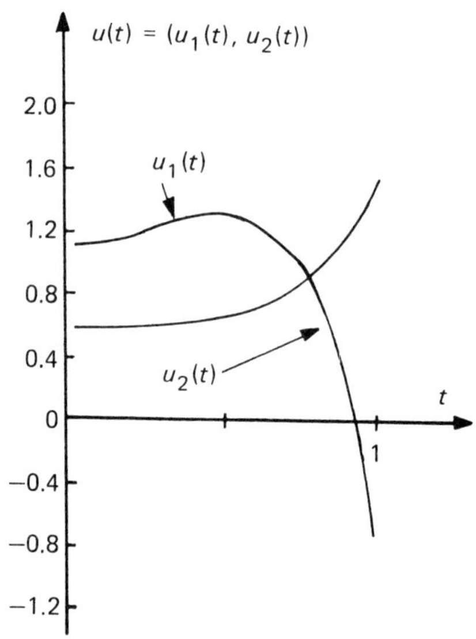

Fig. 3.7 Evolution of the optimal controls

Remark 3.4

The approach considered for the optimal location of controls may also be used for the optimal location of sensors. All the developments are transposable by means of an error criterion for the estimation of the initial state:

$$J = \min_{z_0} \{ \| y(.) - CS(.)z_0 \|^2_{L^2(0,T;\mathbb{R}^q)} \} \tag{3.19}$$

The parameters $\beta_1, \beta_2, \ldots, \beta_m$ which define the support of the sensors yield J as a function to be minimized over the following parameters:

$$\begin{aligned} &\min(J) \\ &\beta_1, \ldots, \beta_m \\ &D_i \subset \Omega \qquad i = 1, \ldots, q \end{aligned} \tag{3.20}$$

3.1.4 Optimum design techniques

In this section, we shall apply to the criterion (3.1) the techniques of optimization of a domain. This approach, which has been developed for a little more than 10 years, furnishes a theoretical method of determining an unknown domain which optimizes some criterion. The method has already been used to determine the optimal design of a heat diffuser in a satellite, of an electromagnet and of a bottle and many other problems in aeronautics.

The difficulties associated with the approach are mainly concerned with the numerical resolution. This is due to the fact that in an iteration, in order to modify the domain, it is necessary to change the whole mesh on the domain. To surmount this difficulty, it is necessary to move the nodes in the interior or else to choose the directions of displacement of the nodes *a priori* and to compute the mesh as a consequence.

Consider the criterion

$$J = \min(\| u \|^2_{L^2(0,T;\mathbb{R}^p)}) = \| u^* \|^2 \tag{3.1}$$

where u^* is given by equation (3.5) is the control which minimizes the energy and assures $z(T) = z_d$. We shall consider the case of one control (Ω_1, g_1). Suppose that D is an open neighbourhood of Ω_1 and let V be deformations defined an $[0, 1] \times D$ such that

$$\begin{aligned} V_1 &= V(.,x(.)) \colon [0,1] \to \mathbb{R}^n \\ V_2 &= V(t,.(t)) \colon D \to \mathbb{R}^n \end{aligned}$$

which is assumed to be sufficiently regular $(V_1 \in C^0, V_2 \in C^k, k \geqslant 1)$. By V, all elements $x_0 \in \Omega_1$ are transformed into $x(t)$, the solution of

$$\begin{aligned} \dot{x}(t) &= V(t, x(t)) \\ x(0) &= x_0 \end{aligned} \tag{3.21}$$

This enables us to define a map $T_t(V)$ which maps the support Ω_1 into Ω_t by

$$\begin{aligned} \Omega_t(V) &= \{x(t) = x(t, x_0) = T_t(V)x_0 \colon x_0 \in \Omega_1\} \\ &= T_t(V)(\Omega_1) \end{aligned} \tag{3.22}$$

As a result, we suppose that the support of the control is varying via the deformation V and derive the value of the criterion J as a function of the new variable Ω_t.

Note that, if the support Ω_1 is connected, then so is Ω_t.

It is necessary to recall the following.

Recall Eulerian derivative

If J is a real-valued function over D, one derives the Eulerian derivative of J at Ω_1

in the direction V and denotes it by $dJ(\Omega_1, V)$, the limit, if it exists:

$$dJ(\Omega_1; V) = \lim_{t\to 0}\left\{\frac{J(\Omega_t) - J(\Omega_1)}{t}\right\} \tag{3.23}$$

For more details, see Zolesio (1979).

Suppose that J is of the form

$$J(\Omega_t) = \int_{\Omega_t} f(\Omega_t, x)\,dx \tag{3.24}$$

$$= \int_{\Omega_1} f(\Omega_t, x) T_t(x) G_t\,dx \tag{3.25}$$

where f is a real-valued function such that the integral exists; G_t is the determinant of the Jacobian of the transformation T_t. It is easy to see that

$$dJ(\Omega_1, V) = \int_{\Omega_1} [\partial_{\Omega_1} f(\Omega_1, x) + f(\Omega, x)\,\mathrm{div}\,\{V(0)\}]\,dx \tag{3.26}$$

Thus, by Stokes' theorem,

$$dJ(\Omega_1, V) = \int_{\Omega_1} \partial_{\Omega_1} f(\Omega_1, x)\,dx + \int_{\partial\Omega_1} f(\Omega_1, \sigma) V(0) n\,d\sigma \tag{3.27}$$

where

$$\partial_{\Omega_1} f(\Omega_1, .) = \lim_{t\to 0}\left\{\frac{f(\Omega_t, .) - f(\Omega_1, .)}{t}\right\}$$

is the derivative of f with respect to the domain Ω_i. ■

We now apply these results to the formulae derived in section 3.1.2. We have

$$M_N(\Omega_1) = \int_0^T S_N(T-\tau) B_N(\Omega_1) B_N^*(\Omega_1) S_N^*(T-\tau)\,d\tau \tag{3.28}$$

and

$$J = J(\Omega_1) = z_{dN}^* M_N^{-1}(\Omega_1) z_{dN}$$

Thus,

$$dJ(\Omega_1, V) = -z_{dN}^* M_N^{-1}(\Omega_1)\,dM_N(\Omega_1, V)\,M_N^{-1}(\Omega_1) z_{dN} \tag{3.29}$$

where

$$dM_N(\Omega_1, V) = \int_0^T S_N(T-\tau)\{dB_N(\Omega_1, V) B_N^*(\Omega_1) + B_N(\Omega_1)\,dB_N^*(\Omega_1, V)] S_N^*(T-\tau)\,d\tau \tag{3.30}$$

When the system is excited by one control, B_N is the column vector

$$B_N = [b_1, b_2, \ldots, b_N]^T$$

with

$$b_i = b_i(\Omega_1) = \langle g_1, \phi_i \rangle_{L^2(\Omega_1)} \qquad 1 \leqslant i \leqslant N$$

Thus,

$$db_i(\Omega, V) = \int_{\Omega_1} \partial g_1(\Omega_1, x)\phi_i(x)\,dx + \int_{\partial\Omega_1} g_1(\Omega, \sigma)\phi_i(\sigma)V(0)n\,d\sigma \tag{3.31}$$

for all $i = 1, \ldots, N$.

So the elements of the matrix $dM_N(\Omega_1, V)$ are

$$(dM_N(\Omega_1, V))_{ij} = \frac{e^{(\lambda_i + \lambda_j)T} - 1}{\lambda_i + \lambda_j}\{db_i(\Omega_1, V)b_j(\Omega_1) + b_i(\Omega_1)\,db_j(\Omega_1, V)\} \tag{3.32}$$

for $i, j = 1, \ldots, N$.

If we let $\hat{Y}_N = (Y_1, \ldots, Y_N)^{\mathrm{T}}$ be the solution of

$$M_N(\Omega_1)\hat{Y}_N = z_{dN} \tag{3.33}$$

(controllability ensures the existence of $\hat{Y}_N$), we obtain

$$\begin{aligned} dJ(\Omega_1, V) = -\sum_{i=1}^{N}\sum_{j=1}^{N} Y_i Y_j \frac{e^{(\lambda_i + \lambda_j)T} - 1}{\lambda_i + \lambda_j}\Bigg[b_j(\Omega_1)\Bigg\{\int_{\Omega_1} \partial_{\Omega_1} g_1(\Omega_1, x)\phi_i\,dx \\ + \int_{\partial\Omega_1} g_1(\Omega_1, \sigma)\phi_i(\sigma)V(0)n\,d\sigma\Bigg\} + b_i(\Omega_i)\Bigg\{\int_{\Omega_1} \partial_{\Omega_1} g_1(\Omega_1, x)\phi_j(x)\,dx \\ + \int_{\partial\Omega_1} g_1(\Omega_1, \sigma)\phi_j(\sigma)V(0)n\,d\sigma\Bigg\}\Bigg] \end{aligned} \tag{3.34}$$

Particular case 1 g_1 independent of Ω_1

It may be that the distribution g is given as function over all Ω and does not depend on the support of the control Ω_1 (for e.g. g constant). Then

$$db_i(\Omega_1, V) = \int_{\partial\Omega_1} g_1(\sigma)\phi_i(\sigma)V(0)n\,d\sigma$$

and so

$$dJ(\Omega_1, V) = -2\sum_{j=1}^{N}\sum_{i=1}^{N} Y_i Y_j \frac{e^{(\lambda_i + \lambda_j)T} - 1}{\lambda_i + \lambda_j} b_i(\Omega_1)\int_{\partial\Omega_1} g_1(\sigma)\phi_j(\sigma)V(0)n\,d\sigma \tag{3.35}$$

Particular case 2 Circular support of the control

This case occurs quite frequently in practice. Suppose that the system is defined on a two-dimensional domain and

$$\Omega_1 = D(x, r), \qquad x = (x_1, x_2)$$

then

$$J(\Omega_1) = J(D(x_1, x_2, r)) \tag{3.36}$$

In this case the deformation V has the components

$$\begin{aligned} V_{x_1}&: D(x_1, x_2, r) \to D(x_1 + t, x_2, r) \\ V_{x_2}&: D(x_1, x_2, r) \to D(x_1, x_2 + t, r) \\ V_r&: D(x_1, x_2, r) \to D(x_1, x_2, r + t) \end{aligned}$$

V_{x_1} and V_{x_2} are translations parallel to the axes and V_r is outwardly normal. The calculation of $\mathrm{d}J(\Omega_1, V)$ requires the partial derivatives of J with respect to the parameters x_1, x_2, r defining the support of the control. These parameters occur in the integral

$$I(x_1, x_2, r) = \int_{\partial\Omega_1} g(\sigma)\phi_j(\sigma)V(0)n\,\mathrm{d}\sigma \tag{3.37}$$

It is easy to see that

$$\frac{\partial I(x_1, x_2, r)}{\partial x_1} = \int_0^{2\pi} g(x_1 + r\cos\theta, x_2 + r\sin\theta)\phi_i(x_1 + r\cos\theta, x_2 + r\sin\theta)\cos\theta\,\mathrm{d}\theta$$

$$\frac{\partial I(x_1, x_2, r)}{\partial x_2} = \int_0^{2\pi} g(x_1 + r\cos\theta, x_2 + r\sin\theta)\phi_i(x_1 + r\cos\theta, x_2 + r\sin\theta)\sin\theta\,\mathrm{d}\theta$$

$$\frac{\partial I(x_1, x_2, r)}{\partial r} = \int_0^{2\pi} g(x_1 + r\cos\theta, x_2 + r\sin\theta)\phi_i(x_1 + r\cos\theta, x_2 + r\sin\theta)\,\mathrm{d}\theta$$

Example 3.2

Let $\Omega =]0, \alpha_1[\times]0, \alpha_2[$ and consider the two-dimensional system defined by

$$\begin{aligned} \frac{\partial y}{\partial t} &= \Delta y + \chi_{\Omega_1}(x)u(t) && \text{in } \Omega \times]0, T[\\ y(x, 0) &= 0 && \text{in } \Omega \\ y(\xi, t) &= 0 && \text{in } \partial\Omega \times]0, T[\end{aligned} \tag{3.38}$$

The system is excited by one control $(\Omega_1, 1)$ and the distribution of the control is independent of Ω_1. Suppose, further, that Ω_1 is circular. Then as a consequence of the theory developed in chapter we know that the control will be strategic if $\alpha_1^2/\alpha_2^2 \notin \mathbb{Q}$ and the coordinates (c_1, c_2) of the centre are such that $c_i/\alpha_i \notin \mathbb{Q}$ for $i = 1$ and 2. The eigenfunctions of the Dirichlet problem are

$$\phi_{mn}(x_1, x_2) = \frac{2}{(\alpha_1\alpha_2)^{1/2}} \sin\left(\frac{m\pi x_1}{\alpha_1}\right) \sin\left(\frac{n\pi x_2}{\alpha_2}\right)$$

and the associated eigenvalues are

$$\lambda_{mn} = -\pi^2\left(\frac{m^2}{\alpha_1^2} + \frac{n^2}{\alpha_2^2}\right)$$

One considers the equivalent system in l^2 and truntcates the system to one of order N. We take as the desired final state

$$z_{\mathrm{d}}(x_1, x_2) = x_1^2 x_2^2(\alpha_1 - x_1)(\alpha_2 - x_2) \tag{3.39}$$

The gradient is calculated from equation (3.34) and the results of the optimization algorithm are summarized in Table 3.2 and Fig. 3.8. Table 3.2 shows the evolution of the parameters which define the support of the control (coordinates of the centre and radius). The cost decreases significantly at first, but subsequent iterations do not have much effect. Fig. 3.8 shows the evolution of the support. The centre tends to the point $(\frac{2}{3}, \frac{2}{3})$ where the desired state has a maximum and the radius evolves in such a way that the support just touches the boundary of the domain.

Table 3.2 Evolution of the parameters defining the support

Iterations	x_1	x_2	r	$J(D(x_1, x_2, r))$
1	0.55	0.63	0.1	1107.74
3	0.65	0.82	0.277	26.51
5	0.650	0.763	0.276	10.048
7	0.646	0.732	0.281	4.505
9	0.639	0.707	0.287	1.735
11	0.634	0.692	0.298	0.741
13	0.630	0.684	0.296	0.382
15	0.625	0.674	0.301	0.083
17	0.6231	0.6691	0.3034	0.0473
19	0.6235	0.6697	0.3029	0.0471
21	0.627	0.6735	0.299	0.0458
23	0.6336	0.6788	0.2929	0.04434
25	0.6346	0.6794	0.2919	0.04409

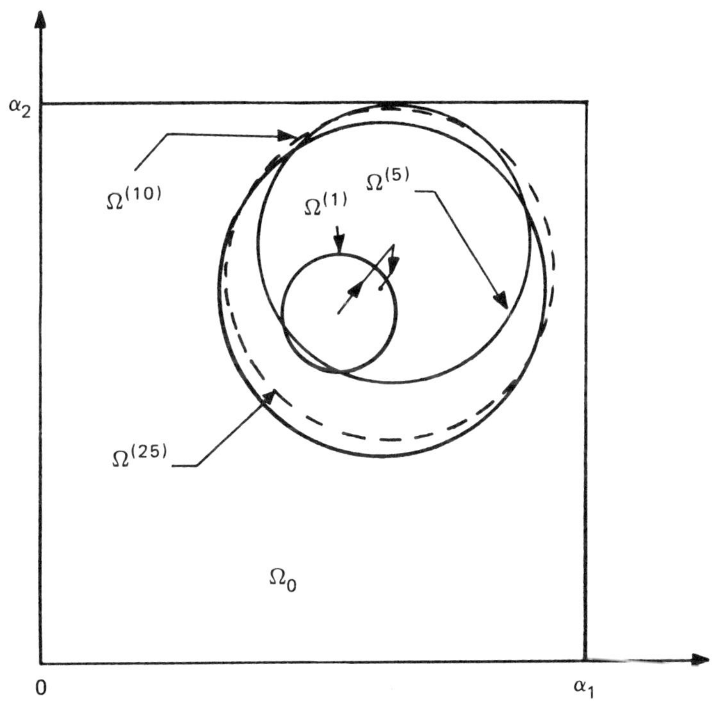

Fig. 3.8 Evolution of the support

3.2 OPTIMAL DISTRIBUTION OF CONTROLS

3.2.1 Introduction

The optimization of the structures which describe the way in which controls enter the system have been the subject of a number of works. We have seen that the location of the supports of controls results in a numerical optimization problem. This is also the case for the distributions over the supports. Many numerical experiments have

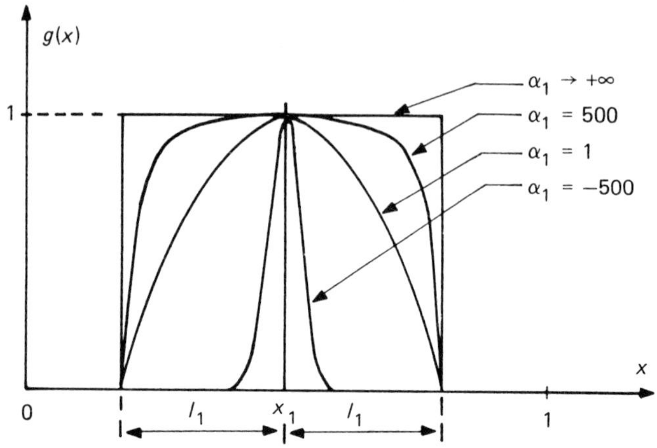

Fig. 3.9

been carried out, particularly for the case of a one-dimensional parabolic system where one takes

$$g(x)=\begin{cases}\dfrac{e^{\alpha_1 l_1^2}}{(1-e^{\alpha_1 l_1^2})}[e^{\alpha_1(x-x_1)^2-\alpha_1 l_1^2}-1] & x_1-l_1\leqslant x\leqslant x_1+l_1\\ 0 & \text{otherwise}\end{cases}\tag{3.40}$$

where x_1 is the centre of the control, α_1 is the curvature of the distribution g and l_1 is the radius of the zone. Thus the determination of control $([x_1-l_1,x_1+l_1],g)$ is accomplished by choice of the parameters (x_1,l_1) for the support and α_1 for the distribution g. The function g in equation (3.40) is regular and may approximate a point control at x_1 as $\alpha_1\to-\infty$ or be constant over the zone $[x_1-l_1,x_1+l_1]$ if $\alpha_1\to+\infty$ (Fig. 3.9).

Numerical experiments for a criterion J of the form in equation (3.1) yield $\alpha_1\to\infty$ and (x_1,l_1) are such that $]x_1-l_1,x_1+l_1[\,=\Omega$, the domain over which the system is defined. In other words, the system should be excited over the largest possible region. Other examples have been given by El Jai (1977, 1978) and El Jai and Najem (1984). These are for simple systems with a few parameters. In the more general case the numerical problems become insurmountable. In the next section, we consider the general case and characterize g as the solution of an optimization problem by introducing it as a supplementary control.

3.2.2 Determination of g as a solution of an optimal control problem

Ω is an open bounded set in $\mathbb{R}^n$ with a regular boundary $\partial\Omega$:

$$\begin{aligned}\frac{\partial y}{\partial t}-\Delta y&=\chi_{\Omega_1}g(x)u(t) && \text{in }\Omega\times]0,T[\\ y(x,0)&=0 && \text{in }\Omega\\ y(\xi,t)&=0 && \text{in }\Omega\times]0,T[\end{aligned}\tag{3.41}$$

with $\Omega_1\subset\Omega$, $g\in L^2(\Omega_1)$ and $u\in L^2(0,T)$. Equations (3.41) represent a parabolic system excited by a zone control (Ω_1,g). We suppose that the support Ω_1 is fixed. Then by

Proposition 1.3, if $\sup_n(r_n) = 1$, there exists a $g \in L^2(\Omega_1)$ such that (Ω_1, g) is strategic. We shall look for a g which is optimal. To do this, we introduce g as a supplementary variable of control. Equations (3.41) admit a unique solution which we denote by $y(x, t; g, u)$. Consider then for $(g, u) \in L^2(\Omega_1) \times L^2(0, T)$ the functional

$$J(g, u) = \| y(., T; g, u) - z_d \|^2_{L^2(\Omega)} + N \| u \|^2_{L^2(0,T)} \tag{3.42}$$

where z_d is given in $L^2(\Omega)$ and $N > 0$. Let K be a convex closed non-empty set in $L^2(\Omega_1)$ and $\mathscr{U}_{ad}$ a convex closed non-empty set in $L^2(0, T)$. We shall consider the following problem (P): determine $(g, u) \in K \times \mathscr{U}_{ad}$ such that

$$J(g, u) \leqslant J(h, v) \qquad \forall (h, v) \in K \times \mathscr{U}_{ad}$$

we have the following result.

Proposition 3.1

For all $g \in L^2(\Omega_1)$, there exists a unique $u_g \in \mathscr{U}_{ad}$ which minimizes $J(g, .)$ on $\mathscr{U}_{ad}$. Furthermore,

$$\langle y(., T; g, v) - y(., T; g, u_g), y(., T; g, u_g) - z_d \rangle + N \langle u_g, v - u_g \rangle \geqslant 0 \tag{3.43}$$

for all $v \in \mathscr{U}_{ad}$.

This is a direct application of the existence and characterization of the minimum of quadratic functionals (Cea, 1971; Lions, 1982).

Proposition 3.2

If K is bounded, then problem (P) admits at least one solution $(g^*, u^*) \in K \times \mathscr{U}_{ad}$ and it is characterized by

$$\begin{aligned} &\langle y(., T; g^*, v) - y(., T; g^*, u^*), y(., T; g^*, u^*) - z_d \rangle + N \langle u^*, v - u^* \rangle \geqslant 0 \\ &\langle y(., T; h, u^*) - y(., T; g^*, u^*), y(., T; g^*, u^*) - z_d \rangle \geqslant 0 \end{aligned} \tag{3.44}$$

for all $v \in \mathscr{U}_{ad}, h \in K$.

Proof Let $(g_n, u_n) \in K \times \mathscr{U}_{ad}$ be a suite such that

$$J(g_n, u_n) \to \inf_{(h,v) \in K \times \mathscr{U}_{ad}} \{J(h, v)\}$$

The suite (g_n) is bounded (since it is contained in K which is bounded) and the suite (u_n) is also bounded (since $J(g_n, u_n)$ is convergent and $\| u_n \|^2 \leqslant (1/N) J(g_n, u_n)$). Hence (g_n, u_n) is bounded in $K \times \mathscr{U}_{ad}$ and so there exists a subsequence (g_{n_k}, u_{n_k}) which converges weakly to (g^*, u^*) in $K \times \mathscr{U}_{ad}$ but, since J is lower semicontinuous in the weak topology,

$$J(g^*, u^*) \leqslant \lim_{n_k \to \infty} \{J(g_{n_k}, u_{n_k}) = \inf_{(h,v) \in K \times \mathscr{U}_{ad}} \{J(h, v)\}$$

For the characterization of (g^*, u^*), u^* minimizes $J(g^*, .)$ on $\mathscr{U}_{ad}$ and hence verifies

$$\langle y(., T; g^*, v) - y(., T; g^*, u^*), y(., T; g^*, u^*) - z_d \rangle + N \langle u^*, v - u^* \rangle \geqslant 0$$

for all $v \in \mathscr{U}_{ad}$.

g^* minimizes $J(.,u^*)$ on K. Let $\tilde{J}(g) = J(g,u^*)$; then

$$\tilde{J}(g) = \pi(g,g) - 2L(g) + \tilde{J}(0)$$

where

$$\pi(g,h) = \langle y(.,T;g,u^*), y(.,T;h,u^*)\rangle$$

is a bilinear symmetric continuous positive form,

$$L(g) = \langle y(.,T;g,u^*), y(.,T;h,u^*)\rangle$$

is a linear continuous form and

$$\tilde{J}(0) = \| z_{\mathrm{d}} \|^2 + N\| u^* \|^2$$

Thus, g^* is characterized by

$$\pi(g^*, h-g^*) - L(h-g^*) \geqslant 0$$

for all $h \in K$, i.e.

$$\langle y(.,T;h,u^*) - y(.,T;g^*,u^*), y(.,T;g^*,u^*) - z_{\mathrm{d}}\rangle \geqslant 0$$

for all $h \in K$. ■

Let us now introduce the adjoint state (which we denote by $p(x,t)$ or $p(x,t;g,u)$) by

$$\begin{aligned} -\frac{\partial p}{\partial t} - \Delta p &= 0 && \text{in } \Omega \times]0,T[\\ p(.,T;g,u) &= y(.,T;g,u) - z_{\mathrm{d}} && \text{in } \Omega \\ p(\xi,t) &= 0 && \text{in } \partial\Omega \times]0,T[\end{aligned} \tag{3.45}$$

Equations (3.45) admit a unique solution. By Green's formula, it is easy to see that

$$\langle y(.,T;g^*,v) - y(.,T;g^*,u^*), p(.,T;g^*,u^*)\rangle = \int_{\Omega_1 \times]0,T[} \{v(t) - u^*(t)\} g^*(x) p(x,t;g^*,u^*)\,\mathrm{d}x\,\mathrm{d}t$$

and, similarly,

$$\langle y(.,T;h,u^*) - y(.,T;g^*,u^*), p(.,T;g^*,u^*)\rangle = \int_{\Omega_1 \times]0,T[} \{h(x) - g^*(x)\} u^*(t) p(x,t;g^*,u^*)\,\mathrm{d}x\,\mathrm{d}t$$

Utilizing these results in equations (3.44), we obtain the optimal conditions which characterize (g^*,u^*) in the form

$$\begin{aligned} &\int_0^T \{v(t) - u^*(t)\}\{Nu^*(t) + \langle g^*, p(.,t;g^*,u^*)\rangle\}\,\mathrm{d}t \geqslant 0 \\ &\int_{\Omega_1} \{h(x) - g^*(x)\}\theta(x)\,\mathrm{d}x \geqslant 0 \end{aligned} \tag{3.46}$$

for all $h \in K$, $v \in \mathscr{U}_{\mathrm{ad}}$, where

$$\theta(.) = \int_0^T p(.,t,g^*,u^*) u(t)\,\mathrm{d}t$$

y and p are the solutions of the equations

$$
\begin{aligned}
&\frac{\partial y}{\partial t} - \Delta y = \chi_{\Omega_1} g u && \text{in } \Omega \times]0, T[\\
&y(z, 0) = 0 && \text{in } \Omega \\
&y(\xi, t) = 0 && \text{in } \partial\Omega \times]0, T[\\
&-\frac{\partial p}{\partial t} - \Delta p = 0 && \text{in } \Omega \times]0, T[\\
&p(., T; g^*, u^*) = y(., T; g^*, u^*) - z_d && \text{in } \Omega \\
&p(\xi, t; g^*, u^*) = 0 && \text{in } \partial\Omega \times]0, T[
\end{aligned}
\tag{3.47}
$$

Proposition 3.3

All pairs (g^*, u^*) which solve problem (P) satisfy equations (3.46) and (3.47).

Remark 3.5 Case of a boundary control

Consider the system excited by a zonal control on the boundary

$$
\begin{aligned}
&\frac{\partial y}{\partial t} - \Delta y = 0 && \text{in } \Omega \times]0, T[\\
&y(x, 0) = 0 && \text{in } \Omega \\
&y(\xi, t) = \chi_{\Gamma_1} g(\xi) u(t) && \text{in } \partial\Omega \times]0, T[
\end{aligned}
\tag{3.48}
$$

with $\Gamma_1 \subset \partial\Omega$, $g \subset L^2(\Gamma_1)$, $u \in L^2(0, T)$. In considering the optimization problem, the first term is taken in $H^{-1}(\Omega)$; then a development similar to that above yields the following.

Proposition 3.4

If K is a closed convex bounded set in $L^2(\Gamma_1)$, $\mathscr{U}_{ad}$ is a closed convex set in $L^2(0, T)$, then the problem (P) admits at least one solution (g^*, u^*) characterized by

$$
\int_0^T \{v(t) - u^*(t)\} \left\{ N u^*(t) - \left\langle g, \frac{\partial p(., t)}{\partial v} \right\rangle \right\} \mathrm{d}t \geqslant 0 \tag{3.49}
$$

$$
\int_{\Gamma_1} \{h(\xi) - g^*(\xi)\} \theta(\xi) \, \mathrm{d}\xi \geqslant 0
$$

for all $v \in \mathscr{U}_{ad}$, $h \in K$, with

$$
\theta(.) = -\int_0^T \frac{\partial p(., t)}{\partial v} u^*(t) \, \mathrm{d}t
$$

y and p are the solutions of

$$
\begin{aligned}
&\frac{\partial y}{\partial t} - \Delta y = 0 && \text{in } \Omega \times]0, T[\\
&y(x, 0) = 0 && \text{in } \Omega \\
&y(\xi, t) = \chi_{\Gamma_1} g(\xi) u(t) && \text{in } \partial\Omega \times]0, T[
\end{aligned}
$$

$$-\frac{\partial p}{\partial t} - \Delta p = 0 \qquad \text{in } \Omega$$
$$p(.,T) = (-\Delta)^{-1}\{y(.,T) - z_{\mathrm{d}}\} \qquad \text{in } \Omega$$
$$p(\xi,t) = 0 \qquad \text{in } \partial\Omega \times]0,T[$$

Other situations (different systems, Neumann boundary conditions, etc.) can also be considered (El Jai and Belfekih, 1985).

3.2.3 Particular case where $K = B(0, M)$

We consider

$$K = \{g \in L^2(\Omega_1) : \|g\|_{L^2(\Omega_1)} \leqslant M, M > 0\}$$
$$\mathcal{U}_{\mathrm{ad}} = L^2(0,T)$$

K is strictly convex. The second equation of equations (3.46) is

$$\langle h - g^*, \theta \rangle \geqslant 0$$

for all $h \in K$. If then $\|\theta\| > 0$, we have

$$g^* = -M\frac{\theta}{\|\theta\|} \tag{3.50}$$

Since $\mathcal{U}_{\mathrm{ad}} = L^2(0,T)$, the first equation of equations (3.46) is

$$Nu^*(t) + \langle p(.,t), g^* \rangle = 0 \tag{3.51}$$

and a direct calculation yields

$$\int_{\Omega_1} \theta(x) g^*(x)\,\mathrm{d}x + N\int_0^T u^{*2}(t)\,\mathrm{d}t = 0 \tag{3.52}$$

Remark 3.6

If $z_{\mathrm{d}} \neq 0$ and $\Omega_1 = \bar{\Omega}$, then $\|\theta\| > 0$. Multiplying equation (3.50) by θ, we have

$$\|\theta\| = -\frac{\langle \theta, g^* \rangle}{M}$$

and hence, from equation (3.52),

$$\|\theta\| = \frac{N}{M}\int_0^T u^{*2}(t)\,\mathrm{d}t$$

Thus,

$$g^* = -\frac{M^2\theta}{N\int_0^T u^{*2}(t)\,\mathrm{d}t}$$

that is to say

$$g^*(x) = -\frac{M^2\int_0^T p(x,t)u^*(t)\,\mathrm{d}t}{N\int_0^T u^{*2}(t)\,\mathrm{d}t} \tag{3.53}$$

3.2.4 Some properties

Consider again the system excited by a zone control (Ω_1, g)

$$\begin{aligned} \frac{\partial y}{\partial t} - \Delta y &= \chi_{\Omega_1} g(x)u(t) && \text{in } \Omega \times]0, T[\\ y(x, 0) &= 0 && \text{in } \Omega \\ y(\xi, t) &= 0 && \text{in } \partial\Omega \times]0, T[\end{aligned} \tag{3.54}$$

The criterion is

$$J(g, u) = \| y(., T; g, u) - z_d \|^2_{L^2(\Omega)} + N \| u \|^2_{L^2(0, T)}$$

with $(g, u) \in L^2(\Omega_1) \times L^2(0, T)$. Let

$$I(g) = \inf_{u \in L^2(0, T)} \{J(g, u)\} = J(g, u^*)$$

The state corresponding to $u = u^*$ is denoted by $y(x, t; g, u^*)$ or $y(x, t; g)$. We have the following properties.

Property 3.1

For all $g \in L^2(\Omega_1)$,

$$I(g) = \| z_d \|^2 - \langle y(., T; g), z_d \rangle$$

This property shows that the state attained at the instant T is always different from z_d since, otherwise, $I(g) = 0$ and then

$$\begin{aligned} I(g) &= \| y(., T; g, u^*) - z_d \|^2 + N \| u^* \|^2 \\ &= N \| u^* \|^2 \end{aligned}$$

So $u^* = 0$, but this implies that $y(., T; g) = 0$ and hence $z_d = 0$.

Proof We have seen in Proposition 3.1 that, for all g, there is a corresponding u_g which minimizes $J(g, .)$. Then,

$$\left. \frac{\partial}{\partial u} J(g, u) \right|_{u = u_g} = 0$$

Hence,

$$\langle y(., T; g), y(., T; g) - z_d \rangle + N \| u_g \|^2 = 0$$

but

$$I(g) = \| y(., T; g) - z_d \|^2 + N \| u_g \|^2$$

from which the result follows. ■

Property 3.2

For all $g \in L^2(\Omega_1)$, we have

$$0 < I(g) \leqslant \| z_d \|^2$$

Proof This result follows from

$$I(g) = \inf_u \{J(g, u)\} \leqslant J(g, u) \qquad \forall u \in L^2(0, T)$$

and, in particular, for $u = 0$. ■

Property 3.3

For all real α such that $|\alpha| \geqslant 1$,

$$I(\alpha g) \leqslant I(g)$$

Proof For all $v \in L^2(0, T)$, we have

$$I(\alpha g) \leqslant J(\alpha g, v)$$

Take $v = (1/\alpha)u$, then

$$I(\alpha g) \leqslant J\left(\alpha g, \frac{u}{\alpha}\right)$$

and

$$\begin{aligned} J\left(\alpha g, \frac{u}{\alpha}\right) &= \left\| y\left(., T; \alpha g, \frac{u}{\alpha}\right) - z_{\mathrm{d}} \right\|^2 + \frac{N}{\alpha^2} \| u \|^2 \\ &= \| y(., T; g, u) - z_{\mathrm{d}} \|^2 + \frac{N}{\alpha^2} \| u \|^2 \\ &\leqslant \| y(., T; g, u) - z_{\mathrm{d}} \|^2 + N \| u \|^2 \\ &= J(g, u) \end{aligned}$$

from which the result follows. ■

Property 3.4

$$\inf_{g \in L^2(\Omega_1)} \{I(g)\} = 0$$

Proof We have seen (Proposition 1.3) that, for a given Ω_1, if $\sup_n (r_n) = 1$, then there always a g such that (Ω_1, g) is strategic. Let g be such a function; then, for all $\varepsilon > 0$, there exists a control $v_\varepsilon \in L^2(0, T)$ such that

$$\| y(., T; g, v_\varepsilon) - z_{\mathrm{d}} \| \leqslant \varepsilon$$

Suppose that $\| z_{\mathrm{d}} \| > \varepsilon$ and consider

$$u_\varepsilon = \varepsilon \frac{v_\varepsilon}{\| v_\varepsilon \|}, \qquad g_\varepsilon = \frac{\| v_\varepsilon \|}{\varepsilon} g$$

Clearly, $g_\varepsilon \in L^2(\Omega_1)$, and

$$\begin{aligned} y(., T; g_\varepsilon, u_\varepsilon) &= y\left(., T; \frac{\| v_\varepsilon \|}{\varepsilon}, \varepsilon \frac{v_\varepsilon}{\| v_\varepsilon \|}\right) \\ &= y(., T; g, v_\varepsilon) \end{aligned}$$

Thus,

$$J(g_\varepsilon, u_\varepsilon) = \| y(.\,, T; g, v_\varepsilon) - z_{\mathrm{d}} \|^2 + N\varepsilon^2$$
$$\leqslant \varepsilon^2(1 + N)$$

and

$$\inf_{g \in L^2(\Omega_1)} \{I(g)\} \leqslant I(g_\varepsilon) \leqslant J(g_\varepsilon, u_\varepsilon) \leqslant \varepsilon^2(1 + N)$$

■

This result shows that we may always choose a spatial distribution to make the cost as small as we like. Clearly, we cannot be sure that this distribution lies in K.

Chapter 4
Hyperbolic and non-linear systems

In this chapter, we shall see that some of the results of the preceding chapters can (under suitable additional assumptions) be extended to other classes of distributed parameter systems.

We shall first consider systems described by hyperbolic partial differential equations and we restrict our main consideration to the controllability problem. For this, we convert the system into the form (*S*) and then apply the results of Chapter 1. The fundamental difference rests in the choice of time parameter T; controllability cannot be achieved unless T is sufficiently large.

Then we examine a class of non-linear systems. Generally, these systems are obtained by linearization but the non-linear term is retained. They can equally be considered as perturbations of the linear system. For the controllability of these systems, it is usually not possible to obtain an explicit control which steers the system to a neighbourhood of the desired state z_{d}. Instead, one generates a control as the limit of a sequence (u_n), the elements of which are determined via the pseudo-inverse technique.

4.1 HYPERBOLIC SYSTEMS

4.1.1 The system

Ω is an open bounded set in $\mathbb{R}^n$ with a smooth boundary $\partial\Omega$ and the system is

$$\frac{\partial^2 y}{\partial t^2} - \Delta y = \sum_{i=1}^{p} g_i u_i \qquad \text{in } \Omega \times]0; T[$$

$$y(x, 0) = \frac{\partial y(x, 0)}{\partial t} = 0 \qquad \text{in } \Omega$$

$$y(\xi, t) = 0 \qquad \text{in } \partial\Omega \times]0, T[$$

The system is excited by p zone controls $(\Omega_i, g_i)_{1\leqslant i\leqslant p}$ with $\Omega_i \subset \Omega$, $g_i \in L^2(\Omega_i)$ for all i, $1 \leqslant i \leqslant p$. As we are interested in the controllability of the system, we shall show how to transform the system into that of the form (S) so that the results of Chapter 1 can be applied:

$$(S) \qquad \begin{cases} \dot{z} = Az + Bu \\ z(0) = 0 \end{cases}$$

4.1.2 Representation by a state vector

In order to transform equation (4.1) into the form (S), let

$$Az = \Delta z, \qquad z \in D(A) = H_0^1(\Omega) \cap H^2(\Omega)$$

The operator $-A$ is self-adjoint and positive and hence we may define

$$M = (-A)^{1/2} \tag{4.2}$$

by

$$\langle Mz, Mz \rangle = -\langle z, Az \rangle$$

Consider now the product space

$$\mathscr{E} = H_0^1(\Omega) \times L^2(\Omega) \tag{4.3}$$

and the operator

$$\bar{A} = \begin{bmatrix} 0 & I \\ A & 0 \end{bmatrix} \tag{4.4}$$

defined by $\bar{A}(z_1, z_2) = (z_2, Az_1)$ for $(z_1, z_2) \in D(\bar{A})$ with

$$D(\bar{A}) = D(A) \times H_0^1(\Omega) \tag{4.5}$$

$\bar{A}$ is a closed linear operator with a dense domain in $\mathscr{E}$ and generates a strongly continuous semigroup $(\mathscr{S}(t))_{t\geqslant 0}$. If (ϕ_{nj}) is a basis of eigenfunctions of Δ for the Dirichlet problem, with associated eigenvalues λ_n of multiplicity r_n, then M has eigenvalues $(-\lambda_n)^{1/2}$. Then it is easy to see that for all $z = (z_1, z_2)$ in $\mathscr{E}$, we have

$$\begin{bmatrix} w_1 \\ w_2 \end{bmatrix} = \mathscr{S}(t) \begin{bmatrix} z_1 \\ z_2 \end{bmatrix} \tag{4.6}$$

with

$$\begin{bmatrix} w_1(t) \\ w_2(t) \end{bmatrix}$$

$$= \begin{bmatrix} \sum_n \sum_{j=1}^{r_n} [\langle z_1, \phi_{n_j} \rangle \cos\{(-\lambda_n)^{1/2} t\} + (-\lambda_n)^{-1/2} \langle z_2, \phi_{nj} \rangle \sin\{(-\lambda_n)^{1/2} t\}] \phi_{nj} \\ \sum_n \sum_{j=1}^{r_n} [-(-\lambda_n)^{1/2} \langle z_1, \phi_{nj} \rangle \sin\{(-\lambda_n)^{1/2} t\} + \langle z_2, \phi_{nj} \rangle \cos\{(-\lambda_n)^{1/2} t\}] \phi_{nj} \end{bmatrix}$$

The scalar product $\langle .,. \rangle$ is in $L^2(\Omega)$. In the particular case where $\Omega =]0, a[$, then

$$\phi_n(x) = \left(\frac{2}{a}\right)^{1/2} \sin\left(\frac{n\pi x}{a}\right) \tag{4.7}$$

and

$$\lambda_n = -\frac{n^2\pi^2}{a^2}$$

For all n, $r_n = 1$, the operator M has eigenvalues $n\pi/a$. In this case the semigroup is given by

$$\begin{bmatrix} w_1(.,t) \\ w_2(.,t) \end{bmatrix} = \begin{bmatrix} \sum_n \left\{ \langle z_1, \phi_n \rangle \cos\left(\frac{n\pi t}{a}\right) + \frac{a}{n\pi} \langle z_2, \phi_n \rangle \sin\left(\frac{n\pi t}{a}\right) \right\} \phi_n(.) \\ \sum_n \left\{ -\frac{n\pi}{a} \langle z_1, \phi_n \rangle \sin\left(\frac{n\pi t}{a}\right) + \langle z_2, \phi_n \rangle \cos\left(\frac{n\pi t}{a}\right) \right\} \phi_n(.) \end{bmatrix} \tag{4.8}$$

Suppose now that $\bar{B}$: $\mathbb{R}^p \to H_0^1(\Omega) \times L^2(\Omega)$ is defined by

$$\bar{B}u = \left[0, \sum_{i=1}^{p} g_i u_i \right] \tag{4.9}$$

for all $u = (u_1, \ldots, u_p) \in \mathbb{R}^p$. Then the equivalent system to the system (4.1) is

$$(S) \quad \begin{cases} \dot{z} = \bar{A}z + \bar{B}u \\ z(0) = 0 \end{cases}$$

where the state is

$$z = \begin{bmatrix} y \\ \dfrac{\partial y}{\partial t} \end{bmatrix} \in L^2(0, T; \mathscr{E})$$

For more details, see Lions (1968) and Curtain and Pritchard (1978).

The analysis of the system (S) requires certain manipulations on the space $\mathscr{E}$. Recall that

(1) $\mathscr{E}$ is a Hilbert space with respect to the inner product

$$\langle z, z' \rangle = \langle Mz_1, Mz_1' \rangle_{L^2(\Omega)} + \langle z_2, z_2' \rangle_{L^2(\Omega)}$$
$$= \int_\Omega [\nabla z_1 \nabla z_1' + z_2 z_2'] \, dx \tag{4.10}$$

where $z = (z_1, z_2)$, $z' = (z_1', z_2') \in \mathscr{E}$;

(2) the adjoint $\bar{A}^*$ of $\bar{A}$ is given by

$$\bar{A}^* = -\bar{A} = \begin{bmatrix} 0 & -I \\ -A & 0 \end{bmatrix} \tag{4.11}$$

and the semigroup adjoint $(\mathscr{S}^*(t))_{t \geqslant 0}$ is given by

$$\mathscr{S}^*(t) = \mathscr{S}(-t) \qquad \text{for all } t \geqslant 0 \tag{4.12}$$

4.1.3 Exact and weak controllability

Exact controllability

Since we have recast the system (4.1) into the form (S), all the definitions in Chapter 1 may be carried over. The characterization of exact controllability (property

1.1) is that (S) is exactly controllable on $[0, T]$ if and only if there exists $\gamma > 0$ such that

$$\| \bar{B}\mathscr{S}^*(.)z^* \|_{L^2(0,T;\mathbb{R}^p)} \geqslant \gamma \| z^* \|_{\mathscr{E}'} \tag{4.13}$$

for all $z \in \mathscr{E}'$.

In general, we do not expect the system to be exactly controllable. However, there are situations where this can be achieved.

Proposition 4.1

The system described by

$$\begin{aligned} \frac{\partial^2 y}{\partial t^2} - \Delta y &= u(x,t) && \text{in } \Omega \times]0,T[\\ y(x,0) &= \frac{\partial y(x,0)}{\partial t} = 0 && \text{in } \Omega \\ y(\xi,t) &= 0 && \text{in } \partial\Omega \times]0,T[\end{aligned} \tag{4.14}$$

is exactly controllable on $[0, T]$ for all $T > 0$.

Proof Using the characterization (4.13) and the expression for the semigroup (4.6), it is easy to see that

$$\left\| \bar{B}^*\mathscr{S}^*(t)\begin{bmatrix} z_1 \\ z_2 \end{bmatrix} \right\|^2 = \sum_n \sum_{j=1}^{r_n} [(-\lambda_n)^{1/2}\langle z_1, \phi_{nj}\rangle \sin\{(-\lambda_n)^{1/2}t\} + \langle z_2, \phi_{nj}\rangle \cos\{(-\lambda_n)^{1/2}t\}]^2$$

Expanding the term in square brackets and calculating the integrals yield

$$\begin{aligned} \left\| \bar{B}^*\mathscr{S}^*(.)\begin{bmatrix} z_1 \\ z_2 \end{bmatrix} \right\|^2_{L^2(0,T;\mathbb{R}^p)} = \frac{1}{2}\sum_n \sum_{j=1}^{r_n} \Bigg\{ & -\lambda_n \langle z_1, \phi_{nj}\rangle^2 \left[T - \frac{\sin\{2(-\lambda_n)^{1/2}T\}}{2(-\lambda_n)^{1/2}} \right] \\ & + 2(-\lambda_n)^{1/2}\langle z_1, \phi_{nj}\rangle\langle z_2, \phi_{nj}\rangle \left[\frac{1-\cos\{2(-\lambda_n)^{1/2}T\}}{2(-\lambda_n)^{1/2}} \right] \\ & + \langle z_2, \phi_{nj}\rangle^2 \left[T + \frac{\sin\{2(-\lambda_n)^{1/2}T\}}{2(-\lambda_n)^{1/2}} \right] \Bigg\} \end{aligned}$$

However, for $z^* = (z_1, z_2)$ in $\mathscr{E}' = \mathscr{E}$,

$$\| z^* \|^2_{\mathscr{E}'} = \sum_n \sum_{j=1}^{r_n} \{ -\lambda_n \langle z_1, \phi_{nj}\rangle^2 + \langle z_2, \phi_{nj}\rangle^2 \}$$

Thus, equation (4.14) is exactly controllable if and only if there exists $\gamma > 0$ such that

$$\begin{aligned} \frac{1}{2}\sum_n \sum_{j=1}^{r_n} & -\lambda_n \langle z_1, \phi_{nj}\rangle^2 \left[T - \frac{\sin\{2(-\lambda_n)^{1/2}T\}}{2(-\lambda_n)^{1/2}} \right] + \langle z_2, \phi_{nj}\rangle^2 \left[T + \frac{\sin\{2(-\lambda_n)^{1/2}T\}}{2(-\lambda_n)^{1/2}} \right] \\ & + 2(-\lambda_n)^{1/2}\langle z_1, \phi_{nj}\rangle\langle z_2, \phi_{nj}\rangle \frac{1-\cos\{2(-\lambda_n)^{1/2}T\}}{2(-\lambda_n)^{1/2}} \\ & \geqslant \gamma^2 \sum_n \sum_{j=1}^{r_n} \{ -\lambda_n \langle z_1, \phi_{nj}\rangle^2 + \langle z_2, \phi_{nj}\rangle^2 \} \end{aligned}$$

This will certainly be the case if there exists $\gamma > 0$ such that

$$X^2\left[T - \frac{\sin\{2(-\lambda_n)^{1/2}T}{2(-\lambda_n)^{1/2}} - 2\gamma^2\right] + 2XY\frac{1-\cos\{2(-\lambda_n)^{1/2}T\}}{2(-\lambda_n)^{1/2}}$$
$$+ Y^2\left[T + \frac{\sin\{2(-\lambda_n)^{1/2}T\}}{2(-\lambda_n)^{1/2}} - 2\gamma^2\right] \geqslant 0$$

for X, $Y \in \mathbb{R}$. Such a γ will exist if

$$T > \left|\frac{\sin\{2(-\lambda_n)^{1/2}T\}}{2(-\lambda_n)^{1/2}}\right|$$

and

$$T^2 - \frac{[\sin\{2(-\lambda_n)^{1/2}T\}]^2}{-4\lambda_n} \geqslant \left[\frac{1-\cos\{2(-\lambda_n)^{1/2}T}{2(-\lambda_n)^{1/2}}\right]^2$$

Both expressions reduce to the requirement

$$T > \left|\frac{\sin\{(-\lambda_n)^{1/2}T\}}{(-\lambda_n)^{1/2}}\right|$$

which is valid for any $T > 0$. ■

This result shows that for all $T > 0$ the system is exactly controllable; this was not the case for parabolic systems (Example 1.1).

Characterization of weak controllability

Again the definition of weak controllability is the same as that in Chapter 1. Consider the linear operator

$$H: u \in L^2(0, T; \mathbb{R}^p) \to Hu = z(T) \in \mathscr{E} \tag{4.15}$$

with

$$z(T) = \left(y(T), \frac{\partial y(T)}{\partial t}\right)$$

Weak controllability is defined by

$$\overline{\mathrm{Im}\,(H)} = H_0^1(\Omega) \times L^2(\Omega) = \mathscr{E} \tag{4.16}$$

The controllability question will depend on the choice of T. We say that the suite of controls $(\Omega_i, g_i)_{1 \leqslant i \leqslant p}$ is strategic in finite time if the system excited by these controls is weakly controllable for some finite time.

Proposition 4.2

The suite of controls $(\Omega_i, g_i)_{1 \leqslant i \leqslant p}$ is strategic in finite time for the hyperbolic system (4.1) if and only if

(1) $p \geqslant \sup_n (r_n)$;

(2) $\mathrm{rank}\,(G_n) = r_n$ for all n, where $(G_n)_{ij} = \langle g_i, \phi_{nj} \rangle_{L^2(\Omega_i)}$, $\quad i = 1, \ldots, p, j = 1, \ldots, r_n$.

Proof The system (4.1) is weakly controllable on $[0, T]$ if and only if, for all

$z=(z_1,z_2)\in\mathscr{E}$,

$$\langle Hu,z\rangle_{\mathscr{E}}=0 \qquad \forall u\in L^2(0,T;\mathbb{R}^p)\Rightarrow z_1=z_2=0$$

but

$$\langle Hu,z\rangle_{\mathscr{E}}=\sum_{i=1}^{p}\sum_{n}\sum_{j=1}^{r_n}\left\{\int_0^T(-\lambda_n)^{1/2}\langle z_1,\phi_{nj}\rangle\langle g_i,\phi_{nj}\rangle\sin\{(-\lambda_n)^{1/2}(T-\tau)\right\}u_i(\tau)\,\mathrm{d}\tau$$
$$+\int_0^T\langle z_2,\phi_{nj}\rangle\langle g_i,\phi_{nj}\rangle\cos\{(-\lambda_n)^{1/2}(T-\tau)\}u_i(\tau)\,\mathrm{d}\tau$$

So $\langle Hu,z\rangle_{\mathscr{E}}=0$ for all $u\in L^2(0,T;\mathbb{R}^p)$ if and only if

$$\sum_{n}\sum_{j=1}^{r_n}(-\lambda_n)^{1/2}\langle z_1,\phi_{nj}\rangle\langle g_i,\phi_{nj}\rangle\sin\{(-\lambda_n)^{1/2}t\}$$
$$+\langle z_2,\phi_{nj}\rangle\langle g_i,\phi_{nj}\rangle\cos\{(-\lambda_n)^{1/2}t\}=0 \tag{4.17}$$

for all $t\in[0,T]$ and i, $1\leqslant i\leqslant p$. Since T may be taken to be arbitrarily large, we see that $\langle Hu,z\rangle_{\mathscr{E}}=0$ if and only if

$$\sum_{j=1}^{r_n}(-\lambda_n)^{1/2}\langle z_1,\phi_{nj}\rangle\langle g_i,\phi_{nj}\rangle=0$$

and

$$\sum_{j=1}^{r_n}\langle z_2,\phi_{nj}\rangle\langle g_i,\phi_{nj}\rangle=0$$

This yields the result. ■

This proposition allows us to carry over to hyperbolic systems all the localization results in Chapter 2.

Example 4.1

Consider $\Omega=]0,a[$; then $r_n=1$ for all n and one control may be sufficient. The system is

$$\begin{aligned}\frac{\partial^2 y}{\partial t^2}&=\frac{\partial^2 y}{\partial z^2}+g_1u_1 &&\text{in }]0,a[\times]0,T[\\ y(x,0)&=\frac{\partial y(x,0)}{\partial t}=0 &&\text{in }]0,a[\\ y(0,t)&=y(a,t)=0 &&\text{in }]0,T[\end{aligned} \tag{4.18}$$

Suppose that the support of the control is $\Omega_1=[x_1-l_1,x_1+l_1]\subset]0,a[$ and $g_1\in L^2(\Omega_1)$. We have the following.

Proposition 4.3

Suppose that $T\geqslant 2a$; then the control (Ω_1,g_1) is strategic on $[0,T]$ for the system (4.17) if and only if

$$\int_{x_1-l_1}^{x_1+l_1}g_1(x)\sin\left(\frac{n\pi x}{a}\right)\mathrm{d}x\neq 0 \tag{4.19}$$

for all $n\geqslant 1$.

Proof In the proof of Proposition 4.2, equation (4.17) for this case is

$$\sum_n \frac{n\pi}{a} \sin\left(\frac{n\pi t}{a}\right)\langle z_1, \phi_n\rangle\langle g_1, \phi_n\rangle + \cos\left(\frac{n\pi t}{a}\right)\langle z_2, \phi_n\rangle\langle g_i, \phi_n\rangle = 0$$

for all $t \in [0, T]$, but the functions

$$\left\{\sin\left(\frac{n\pi t}{a}\right)\right\}_{n \geq 1}, \qquad \left\{\cos\left(\frac{n\pi t}{a}\right)\right\}_{n \geq 0}$$

are a complete orthonormal set in $L^2(0, 2a)$. So we require $\langle g_1, \phi_n\rangle \neq 0$ for all $n \geq 1$. ■

If one considers the case where g_1 is symmetric about $x = x_1$, then the control is strategic if and only if $x_1/a \notin \mathbb{Q}$.

4.1.4 Point controls in]0, *a*[

Consider a system defined on $\Omega =]0, a[$ excited by a single point control located at b, $0 < b < a$:

$$\begin{aligned} &\frac{\partial^2 y}{\partial t^2} - \Delta y = \delta(x - b)u(t) && \text{in }]0, a[\times]0, T[\\ &y(x, 0) = \frac{\partial y(x, 0)}{\partial t} = 0 && \text{in }]0, a[\\ &y(0, t) = y(a, t) = 0 && \text{in }]0, T[\end{aligned} \tag{4.20}$$

One can show by the method of transposition that equation (4.20) admits a unique weak solution. Again we say that the system (4.20) is weakly controllable on $[0, T]$ if

$$\left\{z(T) = \left(y(T), \frac{\partial y(T)}{\partial t}\right) : u \in L^2(0, T)\right\}$$

is dense in $H_0^1(\Omega) \times L^2(\Omega)$. The point control (b, δ_b) is strategic in finite time if there exists a time T such that equation (4.20) is weakly controllable.

Proposition 4.4

Suppose that $T \geq T_0 = 2a$; then the point control (b, δ_b) is strategic on $[0, T]$ for the system (4.20) if and only if $b/a \notin \mathbb{Q}$.

In order to prove this result, we need the following lemma.

Lemma 4.1

For $T = T_0 = 2a$,

$$z(T) = \left(y(T), \frac{\partial y(T)}{\partial t}\right) \in H_0^1(\Omega) \times L^2(\Omega)$$

for all $u \in L^2(0, T_0)$.

Proof Consider a very smooth function h on $[0, T_0]$ such that

$$h(T_0) = h'(T_0) = 0$$

and let $\phi(t) = h(t)\phi_n$, where (ϕ_n) are the eigenfunctions of Δ for the Dirichlet problem. Suppose that λ_n are the corresponding eigenvalues; then if

$$\psi = \phi'' - \Delta\phi$$

we have

$$4\psi = (h'' - \lambda_n h)\phi_n$$

By Green's formula,

$$\int_0^a \int_0^{T_0} 4y \, dt \, dx = \phi_n(b) \int_0^{T_0} u(t)h(t) \, dt$$

but

$$y(.,t) = \sum_n y_n(t)\phi_n(.) \Rightarrow \int_0^{T_0} y_n(h'' - \lambda_n h) \, dt = \phi_n(b) \int_0^{T_0} u(t)h(t) \, dt$$

Since (ϕ_n) are a complete orthonormal set in $L^2(0, T_0)$. A double integration by parts yields

$$\int_0^{T_0} (y_n'' - \lambda_n y_n)h \, dt = \phi_n(b) \int_0^{T_0} u(t)h(t) \, dt$$

Hence

$$\begin{cases} y_n'' - \lambda_n y_n = \phi_n(b)u(t) \\ y_n(0) = y_n'(0) = 0 \end{cases}$$

So

$$y_n(T_0) = \frac{a}{n\pi} \phi_n(b) \int_0^{T_0} \sin\left\{\frac{n\pi(T_0 - \tau)}{a}\right\} u(\tau) \, d\tau$$

$$\frac{\partial y_n(T_0)}{\partial t} = \phi_n(b) \int_0^{T_0} \cos\left\{\frac{n\pi(T_0 - \tau)}{a}\right\} u(\tau) \, d\tau$$

Thus

$$\sum_n \frac{n^2\pi^2}{a^2} [y_n(T_0)]^2 = \sum_n [\phi_n(b)]^2 \left\langle \sin\left\{\frac{n\pi(T_0 - .)}{a}\right\}, u(.)\right\rangle^2_{L^2(0,T_0)} < \infty$$

$$\sum_n \left[\frac{\partial y_n(T_0)}{\partial t}\right]^2 = \sum_n [\phi_n(b)]^2 \left\langle \cos\left\{\frac{n\pi(T_0 - .)}{a}\right\}, u(.)\right\rangle^2_{L^2(0,T_0)} < \infty.$$

Proof of Proposition 4.4 By means of the lemma for $T = T_0 = 2a$ the map

$$H: L^2(0, T) \to H_0^1(\Omega) \times L^2(\Omega)$$

is continuous. The control (b, δ_b) is strategic if and only if $\text{Ker}(H^*) = \{0\}$. It is easy to see that

$$(H^*z)(t) = \sum_n \phi_n(b)\left[\frac{n\pi}{a}\langle z_1, \phi_n\rangle \sin\left\{\frac{n\pi(T_0 - t)}{a}\right\} + \langle z_2, \phi_n\rangle \cos\left\{\frac{n\pi(T_0 - t)}{a}\right\}\right]$$

but

$$\left(\cos\left\{\frac{n\pi(T_0 - .)}{a}\right\}, \sin\left\{\frac{n\pi(T - .)}{a}\right\}\right)$$

are a basis for $L^2(0, T_0)$. So the result follows. ■

4.1.5 Observability

The system (4.1) may be represented in the form

$$(S) \qquad \begin{cases} \dot{z} = \bar{A}z + \bar{B}u \\ z(0) = 0 \end{cases}$$

Consider now the system (S_d)

$$(S_d) \qquad \begin{aligned} \dot{z} &= \tilde{\bar{A}}z, \qquad z(0) = z_0 \\ y_m &= \tilde{\bar{C}}z \end{aligned}$$

We have seen in section 2.5 the following.

The systems (S) and (S_d) are dual one with the other if

$$\tilde{\bar{A}} = \bar{A}^*, \qquad \tilde{\bar{C}} = \bar{B}^*$$

In this case

$$z = \left(y, \frac{\partial y}{\partial t}\right) \text{ and } \bar{B} = \begin{bmatrix} 0 \\ B \end{bmatrix}$$

and so the dual system has $\tilde{\bar{C}} = [0, B^*]$ which means the output equation yields information concerning $\partial y/\partial t$ with

$$y_m = B^* \frac{\partial y}{\partial t}$$

If (S) and (S_d) are dual systems and (S) is exactly or weakly controllable, then (S_d) is exactly or weakly observable. Thus the results obtained for controllability problems can be dualized to obtain observability results. For example, for the system

$$\begin{aligned} &\frac{\partial^2 y}{\partial t^2} - \Delta y = 0 && && \text{in } \Omega \times]0, T[\\ &y(x,0) = y_0, && \frac{\partial y(x,0)}{\partial t} = y_1 && \text{in } \Omega \\ &y(\xi, t) = 0 && && \text{in } \partial\Omega \times]0, T[\end{aligned} \tag{4.21}$$

with output equation

$$y_m(x,t) = \frac{\partial y(x,t)}{\partial t} \qquad \text{in } \Omega \times [0, T]$$

is exactly observable on $[0, T]$ for all $T > 0$. This is immediate by duality with the system (4.14) for which $B = I$ and hence $\tilde{\bar{C}} = [0, I]$. Many other examples can be developed in a similar manner.

4.2 NON-LINEAR SYSTEMS

4.2.1 Introduction

In most real problems the models are non-linear. The study of problems of controllability and observability for such problems is very difficult and usually it is only possible to obtain local results. If one wishes to stabilize a system about some trajectory, then linear models are sufficient. However, if for example one wants to drive the system from some initial state to a desired final state, then the non-linear model must be considered.

In this section, we shall show how a combination of linear theory and fixed point theorems can be used to solve some non-linear problems of controllability, observability and stabilizability.

Consider the system

$$\begin{aligned} \dot{x} &= f(x, u, t) \qquad x(0) = x_0 \\ y &= h(x, u, t) \end{aligned} \tag{4.22}$$

The precise conditions imposed will be made clear later.

The problem of controllability is to drive the system from x_0 to some desired state x_d on $[0, T]$. Whereas the observability problem is to determine the state $x(t)$, $t \geqslant 0$, given the input $u(t)$, $t \geqslant 0$, and the output $y(t)$, $t \in [0, T]$. The approach is the following for both cases. We choose a control $\bar{u}$ and an initial state $\bar{x}_0$ and then solve equation (4.22) to obtain the state $\bar{x}$ and the output $\bar{y}$. Then we have the following.

(1) In the controllability problem, $\bar{x}_0 = x_0$ and $\bar{u}$ is in effect a tentative first choice of control which steers the system to x_d. In general, of course, we do not have $\bar{x}(T) = x_d$.
(2) For the observability problem, we suppose that $\bar{u} = u$ and $\bar{x}_0$ (and so $\bar{x}$) is a first tentative guess for the initial state x_0 (for the state).

Let

$$\begin{aligned} x_0 &= \bar{x}_0 + z_0 \\ x &= \bar{x} + z \\ y &= \bar{y} + \hat{y} \\ u &= \bar{u} + \hat{u} \end{aligned}$$

Then, linearizing about $(\bar{x}, \bar{u}, \bar{y})$, we have

$$\begin{aligned} \dot{z}(t) &= A(t)z(t) + B(t)\hat{u}(t) + N(z(t), \hat{u}(t), t) \\ z(0) &= z_0 \\ \hat{y}(t) &= C(t)z(t) + D(t)\hat{u}(t) + M(z(t), \hat{u}(t), t) \end{aligned} \tag{4.23}$$

where A, B, C and D are linear operators. For simplicity, we assume that they are independent of time.

4.2.2 Controllability

Principal result

We are interested in the problem of choosing $\hat{u}$ such that

$$z(T) = z_{\hat{u}}(T) = x_d - x(T) = z_d$$

Suppose that $z_0 = 0$ and A generates a strongly continuous semigroup $(S(t))_{t \geqslant 0}$. The first equation of equations (4.23) is written in the mild form

$$z(t) = \int_0^t S(t-\tau)B\hat{u}(\tau)\,d\tau + \int_0^t S(t-\tau)N(z(\tau), \hat{u}(\tau), \tau)\,d\tau \tag{4.24}$$

Obviously, we have to impose conditions so that equation (4.24) makes sense. Before we do this, note that, if we are to achieve $z(T)=z_d$, then we require

$$z_d=\int_0^T S(T-\tau)B\hat{u}(\tau)\,d\tau+\int_0^T S(T-\tau)N(z(\tau),\hat{u}(\tau),\tau)\,d\tau \tag{4.25}$$

Consider again the operator H introduced in the previous chapters:

$$H: L^2(0,T;\mathbb{R}^p)\rightarrow\mathscr{E}$$

$$\hat{u}\rightarrow H\hat{u}=\int_0^T S(T-\tau)B\hat{u}(\tau)\,d\tau$$

and denote by $H^\dagger$ the pseudo-inverse. If

$$z_d-\int_0^T S(T-\tau)N(z(\tau),\hat{u}(\tau),\tau)\,d\tau\in \mathrm{Im}\,(H)$$

we may define implicitly a control u^* by

$$u^*=H^\dagger\left[z_d-\int_0^T S(T-\tau)N(z(\tau),u^*(\tau),\tau)\,d\tau\right] \tag{4.26}$$

Substituting in equation (4.23) yields

$$z(t)=\int_0^t S(t-\tau)Bu^*(\tau)\,d\tau+\int_0^t S(t-\tau)N(z(\tau),u^*(\tau),\tau)\,d\tau \tag{4.27}$$

If we can make sense of equations (4.26) and (4.27), then $z(T)=z_d$. So we have reduced the problem of controllability to that of a fixed point for the pair (z,u^*) in equations (4.26) and (4.27).

We now introduce various technical considerations which enable us to solve this problem.

(1) It is necessary to introduce a space in which we seek a fixed point for (z,u^*); we choose

$$\mathscr{X}=C(0,T;\mathscr{E})\times L^2(0,T;\mathbb{R}^p)$$

(2) We define a topology on $\mathrm{Im}\,(H)$ so that it becomes a Banach space which we denote by $\underline{\mathscr{E}}$, then

$$H^\dagger\in\mathscr{L}(\underline{\mathscr{E}},L^2(0,T;\mathbb{R}^p))$$

we suppose also that

$$z_d-\int_0^T S(T-\tau)N(z(\tau),u^*(\tau),\tau)\,d\tau\in\underline{\mathscr{E}}$$

(3) In infinite dimensions, most non-linearities N do not map

$$C(0,T;\mathscr{E})\times L^2(0,T;\mathbb{R}^p)\times[0,T] \qquad \text{into } C(0,T;\mathscr{E})$$

So we assume that N takes its values in

$$L^r(0,T;\overline{\mathscr{E}}) \qquad \text{with } r\geqslant 1$$

where in general $\mathscr{E}\subset\overline{\mathscr{E}}$. We suppose also that the semigroup $(S(t))_{t\geqslant 0}$ is such that

$$\int_0^{\cdot} S(.-\tau)\hat{z}(\tau)\,\mathrm{d}\tau \in C(0,T;\mathscr{E})$$

for all $\hat{z}\in L^r(0,T;\bar{\mathscr{E}})$.

(4) We do not necessarily assume that the operator B is bounded.

With the above assumptions, it is possible to apply a variety of fixed point theorems (Carmichael *et al.*, 1982). One such result is the following.

Theorem 4.1

Suppose there exist constants α, β, γ such that

$$\left\| \int_0^{\cdot} S(.-\tau)Bu(\tau)\,\mathrm{d}\tau \right\|_{C(0,T;\mathscr{E})} \leqslant \alpha \| u \|_{L^2(0,T;\mathbb{R}^p)}$$

$$\left\| \int_0^{T} S(T-\tau)z(\tau)\,\mathrm{d}\tau \right\|_{\mathscr{E}} \leqslant \beta \| z \|_{L^r(0,T;\bar{\mathscr{E}})}$$

$$\left\| \int_0^{\cdot} S(.-\tau)z(\tau)\,\mathrm{d}\tau \right\|_{C(0,T;\mathscr{E})} \leqslant \gamma \| z \|_{L^r(0,T;\bar{\mathscr{E}})}$$

and the application

$$\mathscr{X} \times [0,T] \to L^r(0,T;\bar{\mathscr{E}})$$
$$(z(.),u(.),.) \to N(z(.),u(.),.)$$

is a contraction in a neighbourhood of the origin.

Then there exists a $d\in\mathbb{R}$ such that for all z_{d} with

$$\| z_{\mathrm{d}} \|_{\mathscr{E}} < d$$

equations (4.26) and (4.27) admit a unique fixed point.

The proof of this result utilizes Banach's fixed point (Taylor, 1958; Carmichael *et al.*, 1982). As an illustration, we consider the following simple example.

Example 4.2

Let $\Omega =]0,1[$ and consider the system described by

$$\frac{\partial^2 w}{\partial t^2} = \frac{\partial^2 w}{\partial x^2} + w^2 \qquad \text{in}\,\Omega \times]0,T[$$

$$w(x,0) = \frac{\partial w(x,0)}{\partial t} = 0 \qquad \text{in}\,\Omega \qquad (4.28)$$

$$w(0,t) = u(t), \qquad w(1,t) = 0$$

The state space is taken to be $\mathscr{E} = L^2(\Omega) \times H^{-1}(\Omega)$ and $T = 2$. The semigroup $(S(t))_{t\geqslant 0}$ generated by the operator $\begin{bmatrix} 0 & I \\ \Delta & 0 \end{bmatrix}$ on $\mathscr{E}$ is given by

$$S(t)\begin{bmatrix} z_1 \\ z_2 \end{bmatrix} = \begin{bmatrix} \sum_n [\cos(n\pi t)\langle z_1, \phi_n\rangle + \frac{1}{n\pi}\sin(n\pi t)\langle z_2, \phi_n\rangle]\phi_n \\ \sum_n [-n\pi \sin(n\pi t)\langle z_1, \phi_n\rangle + \cos(n\pi t)\langle z_2, \phi_n\rangle]\phi_n \end{bmatrix}$$

where $\phi_n(x) = 2^{1/2}\sin(n\pi x)$.

Thus

$$\begin{aligned} \| B^* S^*(.)z \|^2_{L^2(0,2)} &= \int_0^2 \left[\sum_n \sin(n\pi t)\langle z_1, \phi_n\rangle + \frac{1}{n\pi}\cos(n\pi t)\langle z_2, \phi_n\rangle \right]^2 \mathrm{d}t \\ &= \sum_n \left\{ \langle z_1, \phi_n\rangle^2 + \frac{1}{n^2\pi^2}\langle z_2, \phi_n\rangle^2 \right\} \\ &= \| z \|^2_{\mathscr{E}} \end{aligned}$$

Thus the linearized system is exactly controllable to $\mathscr{E}$. The non-linearity w^2 is bounded from $L^2(\Omega)$ into $L^1(\Omega)$ and so $\begin{bmatrix} 0 \\ w^2 \end{bmatrix}$ is bounded from $\mathscr{E} \to \mathscr{E}$. It is easy to see that all the conditions of Theorem 4.1 are satisfied and hence the non-linear system (4.28) may be steered to a neighbourhood of the origin.

Algorithms

The fixed point theorem used in Theorem 4.1 immediately suggests an algorithm for the determination of the control. However, a direct application yields an algorithm in terms of the non-linearity N which may be difficult to calculate Carmichael *et al.* (1982), Kassara (1983) and Kassara and El Jai (1983) have developed various algorithms in terms of the given data (4.22) and the linearized operators A and B. Two such algorithms are given below.

Algorithm I

$$\begin{aligned} \dot{x}_{i+1}(t) &= A(t)\{x_{i+1}(t) - x_i(t)\} + f(x_i(t), u_i(t), t) \\ x_{i+1}(0) &= x_0 \\ u_{i+1}(t) &= u_i(t) + H^{\dagger}\{x_{\mathrm{d}} - x_{i+1}(T)\}(t) \end{aligned} \tag{4.29}$$

The starting values are given by

$$\begin{aligned} x_0(t) &= \bar{x}(t) \\ u_0(t) &= \bar{u}(t) \end{aligned}$$

Algorithm II

$$\begin{aligned} \dot{x}_{i+1}(t) &= f(x_{i+1}(t), u_i(t), t) \\ x_i(0) &= x_0 \\ u_{i+1}(t) &= u_i(t) + H^{\dagger}\{x_{\mathrm{d}} - x_{i+1}(T)\}(t) \end{aligned} \tag{4.30}$$

Again the starting values are

$$\begin{aligned} u_0(t) &= \bar{u}(t) \\ x_0(t) &= \bar{x}(t) \end{aligned}$$

The associated linearized system may be exactly controllable in $\underline{\mathscr{E}}$ and, if $x_0 \in \underline{\mathscr{E}}$, the non-linear system may be exactly controllable to a ball $B(x_0, r)$ in $\underline{\mathscr{E}}$ centred at x_0. The radius r of this ball is related to $\| H^\dagger \|$. This suggests that the radius of convergence of the above algorithms may be improved by choosing the parameters of the control (position, distribution, etc.) so that $\| H^\dagger \|$ (or perhaps $\| H^\dagger x_{\mathrm{d}} \|$) is minimized.

4.2.3 Observability

Theoretical approach

The problem of observation of the state is treated in an analogous fashion to the preceding considerations. Again, we simplify by assuming that A and C are time independent. Moreover, we assume that A generates a strongly continuous semigroup $(S(t))_{t \geqslant 0}$ and the pair (A, C) is exactly observable. So there exists $\alpha > 0$ such that

$$\| CS(.)x_0 \|_{\mathscr{Y}} \geqslant \alpha \| x_0 \|_{\mathscr{E}} \tag{4.31}$$

Normally the space of observation is $\mathscr{Y} = L^2(0, T; \theta)$, but this is in general too large or too small to ensure exact observability. So the space $\mathscr{Y}$ is chosen (with in general $\mathscr{Y} \neq L^2(0, T; \theta)$) such that equation (4.31) holds. On the assumption without loss of generality that the input is zero, the state is estimated via the form

$$z(t) = S(t)K^\dagger \left[\hat{y}(.) - M(z(.), .) - C \int_0^{.} S(. - \tau) N(z(\tau), \tau)\, \mathrm{d}\tau \right] + \int_0^t S(t - \tau) N(z(\tau), \tau)\, \mathrm{d}\tau \tag{4.32}$$

where $K^\dagger$ is the pseudo-inverse of the operator K, $(Kx_0)(t) = CS(t)x_0$. We have the following result.

Theorem 4.2

Suppose that there exist α, β and γ such that equation (4.31) is satisfied,

$$\left\| C \int_0^{.} S(. - \tau)\hat{z}(\tau)\, \mathrm{d}\tau \right\|_{\mathscr{Y}} \leqslant \beta \| \hat{z} \|_{L^r(0, T; \bar{\mathscr{E}})}$$

$$\left\| \int_0^{.} S(. - \tau)\hat{z}(\tau)\, \mathrm{d}\tau \right\|_{C(0, T; \mathscr{E})} \leqslant \gamma \| \hat{z} \|_{L^r(0, T; \bar{\mathscr{E}})}$$

and the applications

$$(z(.), .) \rightarrow N(z(.), .)$$
$$C(0, T; \mathscr{E}) \times [0, T] \rightarrow L^r(0, T; \bar{\mathscr{E}})$$

and

$$(z(.), .) \rightarrow M(z(.), .)$$
$$C(0, T; \mathscr{E}) \times [0, T] \rightarrow \mathscr{Y}$$

are contractions on a neighbourhood of the origin.

Then there exists a real d such that for all $\hat{y}$ with

$$\|\hat{y}\|_{\mathscr{Y}} \leqslant d$$

equation (4.32) admits a unique fixed point.

As an application, let us consider the following.

Example 4.3

$\Omega =]0, 1[$, $t \in [0, 2]$ and

$$\begin{aligned}
&\frac{\partial^2 w}{\partial t^2} = \frac{\partial^2 w}{\partial x^2} + w^2 && \text{in } \Omega \times]0, 2[\\
&w(x, 0) = w_1(x), \qquad \frac{\partial w(x, 0)}{\partial t} = w_2(x) && \text{in } \Omega \qquad (4.33) \\
&w(0, t) = w(1, t) = 0 && t \in [0, 2]
\end{aligned}$$

The output equation is taken to be

$$y(t) = \frac{\partial^2 w(0, t)}{\partial x \, \partial t} \qquad t \in [0, 2]$$

The state space $\mathscr{E}$ is taken to be $H_0^1(\Omega) \times L^2(\Omega)$. We have

$$\left\| K \begin{bmatrix} w_1 \\ w_2 \end{bmatrix} \right\|_{L^2(0,2)} = \sum_n [n^4 \pi^4 \langle w_1, \phi_n \rangle^2 + n^2 \pi^2 \langle w_2, \phi_n \rangle^2]$$

So that we do not have exact observability if $\mathscr{Y} = L^2(0, 2)$. We let $\mathscr{Y} = H^{-1}(0, 2)$ where if

$$y(t) = \sum_n \{a_n \sin(n\pi t) + b_n \cos(n\pi t)\}$$

and

$$\| y(.) \|_{\mathscr{Y}}^2 = \sum_n \frac{1}{n^2 \pi^2} (a_n^2 + b_n^2)$$

Then

$$\left\| K \begin{bmatrix} w_1 \\ w_2 \end{bmatrix} \right\|_{\mathscr{Y}}^2 = \left\| \begin{bmatrix} w_1 \\ w_2 \end{bmatrix} \right\|_{\mathscr{E}}^2$$

The non-linearity w^2 is bounded from $H_0^1(\Omega)$ into $L^2(\Omega)$ and so $\begin{bmatrix} 0 \\ w^2 \end{bmatrix}$ is bounded on $\mathscr{E}$. We may therefore apply Theorem 4.2 if $\hat{y}(.) \in H^{-1}(0, 2)$.

Algorithms

Just as in the controllability problem, we may develop various algorithms based on the fixed point theorem used in Theorem 4.2.

Algorithm III

$$\begin{aligned}
&\dot{x}_{i+1}(t) = A(t)\{x_{i+1}(t) - x_i(t)\} + f(x_i(t), t) \\
&x_{i+1}(0) = x_0^i \qquad (4.34) \\
&x_0^{i+1} = x_0^i + K^\dagger[y(.) - h(x_i(.), .) + C(.)\{x_i(.) - x_{i+1}(.)\}]
\end{aligned}$$

Algorithm IV

$$\begin{aligned} \dot{x}_{i+1} &= f(x_{i+1}(t), t), \qquad x_{i+1}(0) = x_0^{i+1} \\ x_0^{i+1} &= x_0^i + K^\dagger\{y(.) - h(x_i(.), .)\} \end{aligned} \tag{4.35}$$

Similar considerations can be made about the structure and location of the sensors. The parameters may be chosen to minimize $\| K^\dagger \|$ (or $\| K^\dagger y \|$).

4.2.4 Compensators

Suppose for simplicity that the non-linear system is

$$\begin{aligned} \dot{z} &= Az + Bu + N(z, u) \\ z(0) &= z_0 \\ y &= Cz + M(z) \end{aligned} \tag{4.36}$$

Consider in $\mathbb{R}^n$ a compensator

$$\begin{aligned} \dot{w} &= \bar{A}w + Ky \\ w(0) &= 0 \\ u &= -Fw \end{aligned} \tag{4.37}$$

We assume that the linear part admits a finite dimensional compensator (see section 1.6) such that

$$\begin{bmatrix} \dot{z} \\ \dot{w} \end{bmatrix} = \begin{bmatrix} A & -BF \\ KC & \bar{A} \end{bmatrix} \begin{bmatrix} z \\ w \end{bmatrix} + \begin{bmatrix} N(z, -Fw) \\ KM(z) \end{bmatrix}$$

where $\begin{bmatrix} A & -BF \\ KC & \bar{A} \end{bmatrix}$ generates an exponentially stable strongly continuous semigroup $(S_e(t))_{t \geqslant 0}$. Let

$$z_e = \begin{bmatrix} z \\ w \end{bmatrix}, \quad y_e = \begin{bmatrix} y \\ w \end{bmatrix}, \quad N_e(z_e) = \begin{bmatrix} N(z, -Fw) \\ KM(z) \end{bmatrix}, \quad M_e(z_e) = \begin{bmatrix} M(z) \\ 0 \end{bmatrix}$$

Then we have

$$\begin{aligned} z_e(t) &= S_e(t) z_e(0) + \int_0^t S_e(t - \iota) N_e(z_e(\iota)) \, d\iota \\ y_e(t) &= C_e z_e(t) + M_e(z_e) \end{aligned} \tag{4.38}$$

Theorem 4.3

Let $E = L^s(0, \infty; \mathscr{E} \times \mathbb{R}^n)$, $s \geqslant 1$, and suppose that there exist two constants α and β such that

$$\| S_e(.) z_{e0} \|_E \leqslant \alpha \| z_{e0} \|_{\mathscr{E} \times \mathbb{R}^n}$$

$$\left\| \int_0^{\cdot} S_e(. - \tau) N_e(z_e(\tau)) \, d\tau \right\|_E \leqslant \beta \| N_e(z_e(.)) \|_{L^r(0, \infty; \bar{\mathscr{E}} \times \mathbb{R}^n)}$$

and the application

$$\begin{gathered} z_e(.) \rightarrow N_e(z_e(.)) \\ L^s(0, \infty; \mathscr{E} \times \mathbb{R}^n) \rightarrow L^r(0, \infty; \bar{\mathscr{E}} \times \mathbb{R}^n) \end{gathered}$$

is a contraction in a neighbourhood of the origin. Then there exists a real d such that for all z_e with

$$\| z_e(0) \|_{\mathscr{E} \times \mathbb{R}^n} < d \tag{4.39}$$

the non-linear system is stable.

The above theorem gives conditions under which the non-linear system can be stabilized. However, in general, the condition (4.39) cannot be checked. Using the developments of section 4.2.3 it is possible to replace this condition by a condition of the form

$$\| y_e \|_{\mathscr{Y}} < d$$

which can be directly verified (Chapman and Pritchard, 1983)

References

Adams, R. A. (1975). *Sobolev spaces*, Academic Press, New York, 1975.

Amouroux, M. (1977). Localisation optimale de captures et d'actionneurs pour la commande d'une classe de systèmes à paramètres répartis. *Thèse d'Etat*, Université Paul Sabatier, Toulouse.

Amouroux, M., and Babary, J. P. (1975). Optimal pointwise control for a class of distributed parameter systems. *Proc. 6th Int. IFAC Congr., Boston Massachusetts, 24–30 August 1975*, International Federation of Automatic Control.

Amouroux, M., and Babary, J. P. (1976). Détermination d'une zône d'action quasi-optimale pour une classe de systémes à paramètres repartis. *C. R. Acad. Soci.*, **282**, pp. 123–126.

Amouroux, M., and Babary, J. P. (1978). On the optimal pointwise control and parametric optimization of distributed parameter systems. *Int. J. Control*, **28**, No. 5, pp. 789–807.

Amouroux, M., and Babary, J. P. (1979). On optimization of zones of action for an optimal control problem for distributed parameter systems, *Int. J. Control*, **29**, No. 5, pp. 861–869.

Amouroux, M., and Babary, J. P. (1982). Selection of Measurement and Control Locations in Distributed Parameter Systems. *Theory and Applications*, (ed. S. G. Tzafestas), Pergamon Press, Oxford, Chapter 10, pp. 295–327.

Amouroux, M., Babary, J. P., and El Jai, A. (1978). Adaptive open loop control for a class of distributed parameter systems. *Automatica*, **14**, pp. 603–607.

Amouroux, M., Courdesses, M., and El Jai, A. (1984). Comparison of two optimal feedback controls for parabolic systems. *Optimal Control Applications and Methods*, Vol. 5, pp. 137–148.

Amouroux, M., Dipillo G., and Grippo, L. (1976). Optimal selection of sensor and control locations for a class of distributed parameter systems. *Ric. Automa.*, **7**, No. 1.

Amouroux, M., El Alami, N., and El Jai, A. (1983). Comment observer l'état d'un système parabolique. *RAIRO, Syst. Anal. Control*, **17**, No. 4, pp. 323–338.

Babary, J. P. (1985). Modelling and adaptive open loop control of tunnel kilns. *Proc. 24th IEEE Cont. on Decision and Control, Fort Lauderdale, December 11–13, 1985*. IEEE, New York.

Babary, J. P., and Souza Leao, J. (1984). Modelling and numerical simulation of an epitaxial reactor. *Math. Comput. Simul.*, **26**, pp. 210–217.

Babary, J. P., and Pelczewski, W. (1985). *Commande Optimale des Systèmes Continus Déterministes*, Masson, Paris.

Balakrishnan, A. V. (1976). *Applied Functional Analysis*, Springer, Berlin.

Banks, H. T., and Kunisch, K. (1982). An approximation theory for nonlinear partial differential

equations with applications to identification and control. *SIAM J. Control Optim.*, **20**.
Barkswill, R. (1983). Algorithm for non-linear observability problems. *M.Sc. Thesis*, University of Warwick.
Benchimol, C. D. (1977). The stabilizability of infinite dimensional time invariant systems. *Ph.D. Thesis*, University of California, Los Angeles.
Ben Israel, A., and Greville, T. N. E. (1973). *Generalized Inverses, Theory and Applications*, Wiley, New York.
Bensoussan, A. (1972). Optimization of sensońs location in a distributed parameter system. *Lect. Notes Math.*, p. 294. 1972.
Berrahmoune, L. (1984). Localisation d'actionneurs pour la contrôlabilité de systèmes paraboliques et hyperboliques. Application par dualité à la localisation de capteurs. *Thèse de Troisi ème Cycle*, Faculté des sciences, Rabat.
Brezis, H. (1983). *Analyse fonctionnelle, Théorie et Applications*, Masson, Paris.
Burger, J. (1976). Emplacement optimal des actionneurs ponctuels ou par zônes dans un systéme gouverné par une équation linéaire d'évolution du premier ordre. *RAIRO J-5*.
Butkovskii, A. G. (1969). *Theory of Optimal Control of Distributed Parameter Systems*, Elsevier, Amsterdam.
Butkovskii, A. G. (1983). *Structural Theory of Distributed Systems*, Ellis Horwood, Chichester, West Sussex.
Butzer P. L., and Berens, H. (1967). *Semi-groups of Operators and Approximation*, Springer, Berlin.
Caravani, P., Dipillo G., and Grippo, L. (1975). Optimal location of a measurement point in a diffusion process. *Proc. 6th Int. IFAC Congr., Boston, Massachusetts, 1985*, International Federation of Automatic Control.
Carmichael, N. Pritchard, A. J., and Quinn, M. D. (1982). State and parameter estimation for nonlinear systems. *Appl. Math. Optim.*, **9**, pp. 133–161.
Cea, J. (1971). *Optimisation, Theorie et Algorithmes*, Dunod, Paris.
Chapman, M. J., and Pritchard, A. J., (1983). Finite dimensional compensators for nonlinear infinite dimensional systems. *Lect. Notes Control Inf. Sci.*, **54**.
Chapman, M. J., Pritchard, A. J., and Verrall, S. (1984). Algorithms for piecewise constant control of nonlinear systems. *Proc. IMA Conf. on Control Theory, Cambridge, 1984*, Institute for Manufacturing Automation.
Chen, W. H. and Seinfeld, J. H. (1975). Oprimal location of process measurements. *Int. J. Control.*, **21**.
Chewning, W. C., and Seidman, T. I. (1977). A convergent scheme for boundary control of the heat equation. *SIAM J. Control* **15**, pp. 64–72.
Courant, R., and Hilbert, D. (1962). *Methods of Mathematical Physics*, Vol. II, Interscience, New York.
Courdesses, M. (1978). Optimal sensors and controllers allocation in output feedback control for a class of distributed parameter systems. *Proc. Symp. on Advances in Measurement and Control, Athens, 1978*, pp. 735–740.
Courdesses, M., Polis, M. P., and Amouroux, M. (1981). On identifiability of parameters in a class of parabolic distributed systems. *IEEE Trans. Autom. Control*, **26**.
Crouch, P. E., Kubrusly, C. S., and Pritchard, A. J. (1982). Systems identification: some fundamental aspects. Rep., No. 104, Control Theory Center, University of Warwick.
Curtain, R. F. (1983). Compensators for infinite dimensional linear systems. *J. Franklin Inst.*, **315**.
Curtain, R. F. (1982). Finite dimensional compensator design for parabolic systems with point and boundary input. IEEE *Trans. Autom. Control*, **26**, pp. 98–104.
Curtain, R. F. (1984). Finite dimensional compensators for parabolic distributed systems with unbounded control and observation. *SIAM J. Control Optim.*, **22**.
Curtain, R. F., Ichikawa, A., and Ryan, E. P. (1976). Optimal location of point sensors. *Proc. IFIP Conf. on Modelling and Identification of Distributed Parameter Systems, Rome, 1976*, International Federation for Information Processing.
Curtain, R. F., and Pritchard, A. J. (1976). A semi-group approach to infinite dimensional systems theory. *Proc. IMA Symp. on Recent Developments in Control, Leicester, 1976*, Institute for Mathematics and its Applications.

Curtain, R. F., and Pritchard, A. J. (1977a). An abstract theory for unbounded control action for distributed parameter systems. *SIAM J. Control*, **15**, pp. 566–611.

Curtain, R. F., and Pritchard, A. J. (1977b). *Functional analysis in Modern Applied Mathematics*, Academic Press, New York.

Curtain, R. F., and Pritchard, A. J. (1978). *Infinite dimensional linear systems theory. Lect. Notes Control Inf. Sci.*, **8.**

Datko, R. (1970). Extending theorem of A. M. Lyapounov to Hilbert space. *J. Math. Anal. Appl.*, **32**, pp. 610–616.

Dautray, R., and Lions, J. L. (1984). *Analyse mathématique et calcul numérique pour les sciences et les techniques. Collection CEA, Série Scientifique*, Masson, Paris.

Delfour, M. C., and Mitter, S. K. (1972). Controllability and observability for infinite dimensional systems. *SIAM J. Control*, **10**, pp. 329–333.

Delfour, M., Payre, G., and Zolesio, J. P. (1982), Optimal design of a minimum weight thermal diffuser with constraint on the output thermal flow. *Appl. Math. Optim.*, **9**, pp. 225–262.

Dolecki, S. (1973). Observability for the one-dimensional heat equation. *Stud. Math.*, **48**, pp. 291–305.

Dolecki, S., and Russell, D. (1977). A general theory of observation and control. *SIAM J. Control.*, **15**, pp. 185–220.

Dunford, N., and Schwartz, J. T. (1963). *Linear Operators*, Interscience, New York.

El Jai, A. (1977). Sur la commande avec estimation de l'état initial d'une classe de systèmes à paramètres répartis. Considérations pratiques. *Rev. RAIRO*, **11**, No. 4, pp. 313–334.

El Jai, A. (1978). Etude d'algorithmes pour la commande de systèmes à paramètres répartis de type parabolique. *Thèse de Doctorat d'Etat*, Université Paul Sabatier, Toulouse.

El Jai, A., and Belfekih, A. (1985). Stratégie optimale pour le contrôle d'un système parabolique. *Rev. Mar. Aut. Inf. Trait. Signal.*

El Jai, A., and Berrahmoune, L. (1983). Localisation d'actionneurs zônes pour la contrôlabilité de systèmes paraboliques. *C. R. Acad. Sci.*, **297**.

El Jai, A., and Berrahmoune. L. (1984a). Localisation d'actionneurs ponctuels pour la contrôlabilité de systèmes paraboliques. *C.R. Acad. Sci.*, **298**, No. 3.

El Jai, A., and Berrahmoune, L. (1984b). Localisation d'actionneurs frontières pour la contrôlabilité de systèmes paraboliques. *C.R. Acad. Sci.*, **298**, No. 8.

El Jai, A., and Najem, A. (1984). Optimal actuator location in a diffusion process. *Lect. Notes Control Inf. Sci.*, **62**, Part 1.

Erdelyi, I. (1972). A generalized inverse for arbitrary operators between Hilbert spaces. *Proc. Cambridge Philos. Soc.*, **71**, Part 1.

Erzberger, H., and Kim, M. (1966). Optimum boundary control of distributed parameter systems. *Int. J. Control*, **9**, pp. 265–278.

Fattorini, H. O. (1966a). On control in finite time of differential equations in Banach space. *Commun. Pure Appl. Math.*, **19**, pp. 17–34.

Fattorini, H. O. (1966b). Some remarks on complete controllability. *SIAM J. Control*, **4**, pp. 686–694.

Fattorini, H. O. (1967a). On complete controllability of linear systems. *J. Differ. Equ.*, **3**, pp. 391–402.

Fattorini, H. O. (1967b). Controllability of higher order linear systems. *Mathematical Theory of Control*, (eds A. V. Balakrishnan and L. W. Neustadt), Academic Press, New York.

Fattorini, H. O. (1975). Boundary control of temperature distributions in a parallelopipedon. *SIAM J. Control*, **13**, pp. 1–13.

Fattorini, H. O. (1976). Local controllability of a nonlinear wave equation. *Math. Syst. Theory*, **9**, pp. 30–45.

Fattorini, H. O., and Russell, D. L. (1971). Exact controllability theorems for linear parabolic equations in one space dimension. *Arch. Ration. Mech. Anal.*, **43**, pp. 272–292.

Faurre, P., and Robin, M. (1984). *Eléments d'Automatique*, Dunod, Paris.

Friedman, A. (1969). *Partial Differential Equations.* Holt, Rinehart and Winston, New York.

Fuhrmann, P. A. (1970). Exact controllability and observability and realization theory in Hilbert space. *J. Math. Anal. Appl.*, **53**, pp. 377–392.

Gibson, J. S. (1980). A note on stabilization of infinite dimensional linear oscillators by compact linear feedback. *SIAM J. Control*, **18**, pp. 311–316.

Glashoff, K., and Weck, N. (1976). Boundary control of parabolic differential equations in arbitrary dimensions: supremum norm problems. *SIAM J. Control*, **14**, pp. 662–681.
Goodson, R. E., and Klein, R. E. (1970). A definition and some results of distributed systems observability. *IEEE Trans. Autom. Control*, **15**.
Goodson, R. E., and Polis, M. P. (1978). Identification of parameters in distributed systems. *Distributed parameter systems* W. H. Ray and D. G. Lainiotis (eds) Dekker, New York.
Goong Chen, Wendell, H., and Crosta, G. (1981). Exact controllability theorems and numerical simulations for some nonlinear differential equations. *SIAM J. Control Optim.*, **19**, No. 6.
Graham, K., and Russell, D. L. (1975). Boundary value control of the wave equation in a spherical region. *SIAM J. Control*, **13**, pp. 174–196.
Hariri, A., and Babary, J. P. (1985). Modelling and simulation of tunnel kilns in ceramic industry. *Int. J. Model. Simul.*, **5**.
Henry, D. (1981). Geometric theory of semi-linear parabolic systems. *Lect. Notes Math.*, **840**.
Hille, E., and Phillips, R. S. (1957). *Functional Analysis and Semi-groups*, American Mathematical Society.
Hinrichsen, D., and Pritchard, A. J. (1986). *Mathematical Systems Theory*, Academic Press, New York.
Ho, L. F., and Russell, D. L. (1983). Admissible input elements for systems in Hilbert space and a Carleson measure criterion. *SIAM J. Control Optim.*, **21**, No. 4, pp. 614–640.
Ichikawa, A., and Ryan, E. P. (1979). Sensor and controller location problems for distributed parameter systems. *Automatica*, **15**.
Kailath, T. (1980). *Linear Systems*, Prentice-Hall, Englewood Cliffs, New Jersey.
Kappel, F., and Schappacher, W. (1981). *Abstract Cauchy Problems and Functional Differential Equations, Research Notes in Mathematics*, No. 48, Pitman, London.
Kappel, F., Kunisch, K., and Schappacher, W. (1983). Control theory for distributed parameter systems and applications. *Lect. Notes Control Inf. Sci.*, **54**.
Kassara, K. (1983). Commandabilité et observabilité d'une class de systèmes non linéaires distribués. Approche théorique et algorithmes. *Thèse*, Faculté des Sciences, Rabat.
Kassara, K., and El Jai, A. (1983). Algorithmes pour la commande d'une classe de systèmes à paramètres répartis non linéaires. *Rev. Mar. Aut. Inf. Trait. Signal*, **1**, No. 1, pp. 3–24.
Kato, T. (1980). *Perturbation Theory for Linear Operators*, 2nd edn, Springer, Berlin.
Kitamura, S., and Nakagiri, S. (1977). Identifiability of spatially varying and constant parameters in distributed systems of parabolic type. *SIAM J. Control Optim.*, **15**.
Kobayashi, T. (1976). Initial state determination for distributed parameter systems. *SIAM J. Control.*, **14**, pp. 934–944.
Kubrusly, C. S. (1977). Distributed parameter system identification: a survey. *Int. J. Control*, **26**.
Kubrusly, C. S., and Malebranche, H. (1985). Sensors and controllers location in distributed systems: a survey. *Automatica*, **21**, No. 2, pp. 117–128.
Kumar, S., and Seinfeld, J. H. (1978). Optimal location of measurements in tubular reactors. *Chem. Eng. Sci.*, **33**.
Kwakernaak, H., and Sivan, R. (1972). *Linear Optimal Control Systems*, Wiley, New York.
Lamont, G. B., and Kumar, K. S. P. (1972). State estimation in distributed parameter systems via least square and invariant embedding. *J. Math. Anal. Appl.*, **38**, pp. 588–606.
Leigh, J. R. (1980). *Functional Analysis and Linear Control Theory*, Academic Press, New York.
Le Pourhiet, A., and Le Letty, L. (1976). Optimization of sensors location in distributed parameter identification. *Proc. 4th IFAC Symp. on Identification and System Parameter Estimation, Tbilisi, 1976*, International Federation of Automatic Control, pp. 1581–1591.
Li Ta Tsien (1979). Propriétés d'espaces fonctionnels intervenant en contrôle optimal. *C.R. Acad. Sci.*, **298**, p. 687.
Lions, J. L. (1968). *Contrôle Optimal des Systèmes Gouvérnés par des Équations aux Dérivées Partielles*, Dunod, Paris.
Lions, J. L. (1969). *Quelques Méthodes de Résolution de Problèmes aux Limites non Linéaires*, Dunod, Paris.
Lions, J. L. (1980a). On the foundation of the optimal control of distributed systems. *Lect. Notes Control Inf. Sci.*, **22**.
Lions, J. L. (1980b). Function spaces and optimal control of distributed systems. *Notes*, Instituto de Matematico, Universidade Federal de Rio de Janeiro.

Lions, J. L. (1981). *Some Methods in the Mathematical Analysis of Systems and their Control*, Science Press, Beijing.

Lions, J. L. (1982). *Contrôle de Système Distribués Singuliers, Série MMI*, Dunod, Paris.

Lions, J. L., Personnel notes, 1986.

Lions, J. L., and Magenes, E. (1968). *Problèmes aux Limites non Homogènes et Applications*, Vols 1 and 2, Dunod, Paris.

Magnusson, K., Pritchard, A. J., and Quinn, M. D. (1980). The application of fixed point theorems to global non-linear controllability problems. *Rep.*, No. 87, Control Theory Centre, University of Warwick.

Manitius, A., and Triggiani, R. (1976). Controllability, observability and stabilizability of retarded systems. *Proc. IEEE Conf. on Decision and Control*, IEEE, New York.

Martin, J. Cl. E. (1976). On the controllability of parabolic systems by scanning controls. *IEEE Trans. Autom. Control*, **21**, No. 4.

Najim, M., Najim, K., and Zyoute, M. (1976). Modelling and identification of a phosphate furnace. *Proc. IFAC Symp. on Identification and System Parameter Estimation, Tbilisi, 1976*, International Federation of Automatic Control.

Nambu, T. (1984). On the stabilization of diffusion equations: boundary observation and feedback. *J. Differ. Equ.*, **52**.

Omatu, S., Koide, S., and Soeda, T. (1978). Optimal sensor location problem for linear distributed parameter systems. *IEEE Trans. Autum. Control*, **23**.

Omatu, S., and Seinfeld, J. H. (1983). Optimization of sensor and actuator location in a distributed parameter system. *J. Franklin Inst.*, **315**.

Pironneau, O. (1983). Optimal shape design for elliptic systems. *Lect. Notes Comput. Phys.*

Polis, M. P. (1982). The distributed system parameter identification problem: a survey of recent results. *Proc. 3rd IFAC Symp. on Control of Distributed Parameter Systems, Toulouse, 1982*, International Federation of Automatic Control.

Pourciel, J. B., Babary, J. P., and Barreteau, D. (1984). Sur l'exploitation des mesures de temperature dans an four de tuilerie industrielle. *Congr. Mesures thermiques, application aux systèms et aux processus industriels: diagnostic, contrôle et conduite, Toulouse, 28–30 May 1984*.

Pritchard, A. J. (1969). Stability and control of distributed systems. *Proc. IEEE*, pp. 1433–1438.

Pritchard, A. J. (1982). Nonlinear infinite dimensional systems. *Proc. 3rd IFAC Symp. on Control of Distributed Parameter Systems, Toulse, 1982.* International Federation of Automatic Control.

Pritchard, A. J., and Triggiani, R. (1974). Stabilizability in Banach spaces. *Rep.*, No. 35, Control Theory Centre, University of Warwick.

Pritchard, A. J., and Wirth, A. (1978). Unbounded control and observation systems and their duality. *SIAM J. Control Optim.*, **16**, No. 4, pp. 535–545.

Pritchard, A. J., and Zabczyk, J. (1977). Stability and stabilizability of infinite dimensional systems. Rep. No. 70, Control Theory Centre, University of Warwick.

Ray, W. H. (1978). Some recent applications of distributed parameter systems theory. A survey. *Automatica*, **14**.

Ray, W. H., and Lainiotis, D. G. (1978). *Distributed parameter Systems—Identification, Estimation and Control*, Dekker, New York.

Redouane, L. (1980). Commande sous-optimale d'un four industriel de séchage de phosphates par microprocesseur. *Thèse de Troisième Cycle*, Faculté des Sciences, Rabat.

Robinson, A. C. (1971). A survey of optimal control of distributed parameter systems. *Automatica.*, **7**.

Ruberti, A. (1978). *Distributed parameter systems: modelling and identification. Lect. Notes Control Inf. Sci.*, **1**.

Russek, A. (1975). On Relations between Exact Controllability and Rank Condition, Preprint, Institute of Mathematics, Warsaw.

Russell, D. L. (1967). On boundary value controllability of linear symmetric hyperbolic systems. *Mathematical Theory of Control* (eds A. V. Balakrishnan and L. W. Neustadt) Academic Press, New York.

Russell, D. L. (1969). Linear stabilization of the linear oscillator in Hilbert space. *J. Math. Anal. Appl.*, **25**, pp. 663–675.

Russell, D. L. (1972). Control theory of hyperbolic equations related to certain questions in harmonic analysis and spectral theory. *J. Math. Anal. Appl.*, **40**, pp. 336–368.

Russell, D. L. (1975). Problems of control and stabilization for partial differential equations. *Proc. 6th Int. IAFC Congr., Boston Massachusetts. 24–30 August 1975*, International Federation of Automatic Control.

Russell, D. L. (1978). Controllability and stabilizability theory for linear partial differential equations: recent progress and open questions. *SIAM Rev.*, **20**, pp. 639–739.

Sakawa, Y. (1974). Controllability for partial differential equations of parabolic type. *SIAM J. Control*, **12**, pp. 389–400.

Sakawa, Y. (1975). Observability and related problems for partial differential equations of parabolic type. *SIAM J. Control*, **13**, pp. 14–27.

Salamon, D. (1984). *Control and Observations of Neutral Systems, Research Notes in Mathematics*, No. 91, Pitman, London.

Sawaragi, Y., Soeda, T., and Omatu, S. (1978). Modelling, estimation and their applications for distributed parameter systems. *Let. Notes Control Inf. Sci.*, **11**.

Schumacher, J. M. (1981). *Dynamic Feedback in Finite and Infinite Dimensional Linear Systems, Mathematical Center Tracts*, No. 143, Mathematish Centrum, Amsterdam.

Seidman, T. I. (1973). Problems of boundary control and observation for diffusion processes. *Math. Res. Rep.*, No. 73–10, University of Maryland.

Seidman, T. I. (1977). Exact boundary control for some evolution equations. *Math. Res. Rep.*, No. 77–4, University of Maryland.

Slemrod, M. (1974a). A note on complete controllability and stabilizability of linear control systems in Hilbert space. *SIAM J. Control*, **12**, pp. 500–508.

Slemrod, M. (1974b). The linear stabilization problem in Hilbert space. *J. Funct. Anal.*, **11**, pp. 334–345.

Souza Leao, J., Gouyon, J. P., and Babary, J. P. (1983). Numerical simulation of the control of the space–time thermal profile profile in an epitaxial reactor. *Proc. Int. Symp. IMACS on Simulation in Engineering Sciences, Nantes, 9–11 May 1983*, International Association for Mathematics and Computers in Simulation.

Szego, G. (1959). *Orthogonal Polynomials, Colloquium publications*, American Mathematical Society, Providence, Rhode Island.

Taylor, A. E. (1958). *Introduction to Functional Analysis*, Wiley, New York.

Triggiani, R. (1975a). On the lack of exact controllability for mild solutions in Banach spaces. *J. Math. Anal. Appl.*, **50**, pp. 438–446.

Triggiani, R. (1975b). Controllability and observability in Banach spaces with bounded operators. *SIAM J. Control*, **13**.

Triggiani, R. (1976). Extension of rank conditions for controllability and observability to Banach spaces and unbounded operators. *SIAM J. Control Optim.*, **14**, pp. 313–338.

Triggiani, R. (1977). A note on the lack of exact controllability for mild solutions in Banach spaces. *SIAM J. Control Optim.*, **15**, pp. 407–411.

Triggiani, R. (1978). On the relationship between first and second order controllable systems in Banach spaces. *SIAM J. Control Optim.*, **16**, pp. 847–859.

Tzafertas, S. G. (1978). Distributed parameter state estimation. *Distributed Parameter Systems*, (eds W. H. Ray and D. G. Lainiotis) Dekker, New York.

Vidyasagar, M., and Higgins, T. J. (1973). A basic theorem on distributed control and point control. *J. Dyn. Syst., Meas. Control*, **67**.

Wang, P. K. C. (1971). Feedback stabilization of distributive systems with application to plasma stabilization. *Instability of Continuous Systems*, (ed. H. Leipholz), Springer, Berlin.

Washburn, D. C. (1976). A semi-group theoretic approach to modelling of boundary input problems. *Proc. IFIP Conf., Rome, 1976*, International Federation for Information Processing.

Weidmann, J. (1980). *Linear Operators in Hilbert Spaces*, Springer, Berlin.

Yosida, K. (1980). *Functional Analysis*, 6th edn, Springer, Berlin.

Yu, T. K., and Seinfeld, J. H. (1971). Observability of a class of hyperbolic distributed parameter systems. *IEEE Trans. Autom. Control*, **16**.

Yu, T. K. and Seinfeld, J. H. (1973). Observability and optimal measurement location in linear distributed parameter systems. *Int. J. Control*, **18**.

Zabczyk, J. (1976). Complete stabilizability implies exact controllability. *Seminarul de Ecuatti Functionale, Universitatea din Timosoara, Rumania, 1976.*
Zabczyk, J. (1977). A semi-group approach to boundary value control. *Proc. 2nd IFAC Symp. on Control of Distributed Parameter Systems, Warwick, 1977*, International Federation of Automatic Control.
Zolesio, J. P. (1979). Identification de domaines par déformations. *Thèse d'Etat*, Nice.
Zolesio, J. P. (1980). Domain variational formulation for free boundary problems. *Optimization of Distributed Parameter Structures*, (eds E. J. Haug and J. Cea), Sijthoff and Noordhoff, Alphen aan den Rihn, 1980.

Index

Mathematics and its Applications

Series Editor: G. M. BELL, Professor of Mathematics, King's College London (KQC), University of London

Faux, I.D. & Pratt, M.J.	**Computational Geometry for Design and Manufacture**
Firby, P.A. & Gardiner, C.F.	**Surface Topology**
Gardiner, C.F.	**Modern Algebra**
Gardiner, C.F.	**Algebraic Structures: with Applications**
Gasson, P.C.	**Geometry of Spatial Forms**
Goodbody, A.M.	**Cartesian Tensors**
Goult, R.J.	**Applied Linear Algebra**
Graham, A.	**Kronecker Products and Matrix Calculus: with Applications**
Graham, A.	**Matrix Theory and Applications for Engineers and Mathematicians**
Graham, A.	**Nonnegative Matrices and Applicable Topics in Linear Algebra**
Griffel, D.H.	**Applied Functional Analysis**
Griffel, D.H.	**Linear Algebra**
Hanyga, A.	**Mathematical Theory of Non-linear Elasticity**
Harris, D.J.	**Mathematics for Business, Management and Economics**
Hoskins, R.F.	**Generalised Functions**
Hoskins, R.F.	**Standard and Non-standard Analysis**
Hunter, S.C.	**Mechanics of Continuous Media, 2nd (Revised) Edition**
Huntley, I. & Johnson, R.M.	**Linear and Nonlinear Differential Equations**
Jaswon, M.A. & Rose, M.A.	**Crystal Symmetry: The Theory of Colour Crystallography**
Johnson, R.M.	**Theory and Applications of Linear Differential and Difference Equations**
Johnson, R.M.	**Calculus: Theory and Applications in Technology and the Physical and Life Sciences**
Jones, R.H. & Steele, N.C.	**Mathematics of Communication**
Kelly, J.C.	**Abstract Algebra**
Kim, K.H. & Roush, F.W.	**Applied Abstract Algebra**
Kim, K.H. & Roush, F.W.	**Team Theory**
Kosinski, W.	**Field Singularities and Wave Analysis in Continuum Mechanics**
Krishnamurthy, V.	**Combinatorics: Theory and Applications**
Lindfield, G. & Penny, J.E.T.	**Microcomputers in Numerical Analysis**
Lord, E.A. & Wilson, C.B.	**The Mathematical Description of Shape and Form**
Malik, M., Riznichenko, G.Y. & Rubin, A.B.	**Biological Electron Transport Processes and their Computer Simulation**
Marichev, O.I.	**Integral Transforms of Higher Transcendental Functions**
Massey, B.S.	**Measures in Science and Engineering**
Meek, B.L. & Fairthorne, S.	**Using Computers**
Mikolas, M.	**Real Functions and Orthogonal Series**
Moore, R.	**Computational Functional Analysis**
Müller-Pfeiffer, E.	**Spectral Theory of Ordinary Differential Operators**
Murphy, J.A. & McShane, B.	**Computation in Numerical Analysis**
Nonweiler, T.R.F.	**Computational Mathematics: An Introduction to Numerical Approximation**
Ogden, R.W.	**Non-linear Elastic Deformations**
Oldknow, A.	**Microcomputers in Geometry**
Oldknow, A. & Smith, D.	**Learning Mathematics with Micros**
O'Neill, M.E. & Chorlton, F.	**Ideal and Incompressible Fluid Dynamics**
O'Neill, M.E. & Chorlton, F.	**Viscous and Compressible Fluid Dynamics**
Page, S. G.	**Mathematics: A Second Start**
Prior, D. & Moscardini, A.O.	**Model Formulation Analysis**
Rankin, R.A.	**Modular Forms**
Scorer, R.S.	**Environmental Aerodynamics**
Smith, D.K.	**Network Optimisation Practice: A Computational Guide**
Srivastava, H.M. & Karlsson, P.W.	**Multiple Gaussian Hypergeometric Series**
Srivastava, H.M. & Manocha, H.L.	**A Treatise on Generating Functions**
Shivamoggi, B.K.	**Stability of Parallel Gas Flows**
Stirling, D.S.G.	**Mathematical Analysis**
Sweet, M.V.	**Algebra, Geometry and Trigonometry in Science, Engineering and Mathematics**
Temperley, H.N.V. & Trevena, D.H.	**Liquids and Their Properties**
Temperley, H.N.V.	**Graph Theory and Applications**
Thom, R.	**Mathematical Models of Morphogenesis**
Toth, G.	**Harmonic and Minimal Maps**
Townend, M. S.	**Mathematics in Sport**
Towend, M.S. & Pountney, D.C.	**Computer-aided Engineering Mathematics**
Twizell, E.H.	**Computational Methods for Partial Differential Equations**
Twizell, E.H.	**Numerical Methods in the Biomedical Sciences**
Vince, A. and Morris, C.	**Mathematics for Computer Studies**
Walton, K., Marshall, J., Gorecki, H. & Korytowski, A.	**Control Theory for Time Delay Systems**
Wheeler, R.F.	**Rethinking Mathematical Concepts**
Willmore, T.J.	**Total Curvature in Riemannian Geometry**
Willmore, T.J. & Hitchin, N.	**Global Riemannian Geometry**

Statistics and Operational Research

Editor: B. W. CONOLLY, Professor of Operational Research, Queen Mary College, University of London

Beaumont, G.P.	**Introductory Applied Probability**
Beaumont, G.P.	**Probability and Random Variables**
Conolly, B.W.	**Techniques in Operational Research: Vol. 1, Queueing Systems**
Conolly, B.W.	**Techniques in Operational Research: Vol. 2, Models, Search, Randomization**
Conolly, B.W.	**Lecture Notes in Queueing Systems**
French, S.	**Sequencing and Scheduling: Mathematics of the Job Shop**
French, S.	**Decision Theory: An Introduction to the Mathematics of Rationality**
Griffiths, P. & Hill, I.D.	**Applied Statistics Algorithms**
Hartley, R.	**Linear and Non-linear Programming**
Jolliffe, F.R.	**Survey Design and Analysis**
Jones, A.J.	**Game Theory**
Kemp, K.W.	**Dice, Data and Decisions: Introductory Statistics**
Moscardini, A.O. & Robson, E.H.	**Mathematical Modelling for Information Technology**
Oliveira-Pinto, F.	**Simulation Concepts in Mathematical Modelling**
Oliveira-Pinto, F. & Conolly, B.W.	**Applicable Mathematics of Non-physical Phenomena**
Ratschek, J. & Rokne, J.	**Computer Methods for the Range of Functions**
Ratschek, J. & Rokne, J.	**New Computer Methods for Global Optimisation**
Schendel, U.	**Introduction to Numerical Methods for Parallel Computers**
Stoodley, K.D.C.	**Applied and Computational Statistics: A First Course**
Stoodley, K.D.C., Lewis, T. & Stainton, C.L.S.	**Applied Statistical Techniques**
Thomas, L.C.	**Games, Theory and Applications**
Whitehead, J.R.	**The Design and Analysis of Sequential Clinical Trials**